Bhavian Patel

Effects of testosterone and anabolic steroids on muscle hypertrophy

Bhavian Patel

Effects of testosterone and anabolic steroids on muscle hypertrophy

Physiological mechanisms involved in skeletal muscle hypertrophy

LAP LAMBERT Academic Publishing

Impressum / Imprint
Bibliografische Information der Deutschen Nationalbibliothek: Die Deutsche Nationalbibliothek verzeichnet diese Publikation in der Deutschen Nationalbibliografie; detaillierte bibliografische Daten sind im Internet über http://dnb.d-nb.de abrufbar.
Alle in diesem Buch genannten Marken und Produktnamen unterliegen warenzeichen-, marken- oder patentrechtlichem Schutz bzw. sind Warenzeichen oder eingetragene Warenzeichen der jeweiligen Inhaber. Die Wiedergabe von Marken, Produktnamen, Gebrauchsnamen, Handelsnamen, Warenbezeichnungen u.s.w. in diesem Werk berechtigt auch ohne besondere Kennzeichnung nicht zu der Annahme, dass solche Namen im Sinne der Warenzeichen- und Markenschutzgesetzgebung als frei zu betrachten wären und daher von jedermann benutzt werden dürften.

Bibliographic information published by the Deutsche Nationalbibliothek: The Deutsche Nationalbibliothek lists this publication in the Deutsche Nationalbibliografie; detailed bibliographic data are available in the Internet at http://dnb.d-nb.de.
Any brand names and product names mentioned in this book are subject to trademark, brand or patent protection and are trademarks or registered trademarks of their respective holders. The use of brand names, product names, common names, trade names, product descriptions etc. even without a particular marking in this work is in no way to be construed to mean that such names may be regarded as unrestricted in respect of trademark and brand protection legislation and could thus be used by anyone.

Coverbild / Cover image: www.ingimage.com

Verlag / Publisher:
LAP LAMBERT Academic Publishing
ist ein Imprint der / is a trademark of
OmniScriptum GmbH & Co. KG
Heinrich-Böcking-Str. 6-8, 66121 Saarbrücken, Deutschland / Germany
Email: info@lap-publishing.com

Herstellung: siehe letzte Seite /
Printed at: see last page
ISBN: 978-3-659-77794-3

Zugl. / Approved by: Leicester, University of Leicester, Dissertation, 2015

Copyright © 2015 OmniScriptum GmbH & Co. KG
Alle Rechte vorbehalten. / All rights reserved. Saarbrücken 2015

Contents

Abstract 3
1. Introduction 5
2. Skeletal muscle hypertrophy 8
    2.1 Muscle fibre structure 8
    2.2 Muscle fibre types 9
    2.3 Muscle hypertrophy 10
    2.4 Additional factors affecting muscle growth; nutrition and rest 12
    2.5 Motor neurons and neuronal involvement in muscle contraction 13
3. Testosterone 18
    3.1 Testosterone and its synthetic derivatives 18
    3.2 Testosterone mechanisms of action 21
    3.3 Testosterone effects on skeletal muscle fibres 22
    3.4 The action of testosterone on myonuclei and satellite cells 24
    3.5 Androgen Receptor (AR) expression 25
    3.6 Hormonal signalling mechanisms and endogenous testosterone
        production 27
    3.7 Possible non-genomic mode of action of testosterone 29
4. Anabolic Androgenic Steroids (AAS) 32
    4.1 Exogenous testosterone administration 32
    4.2 Anabolic effects of testosterone and AAS synthetic derivatives 36
5. Side effects of AAS use 43
    5.1 Cardiovascular effects 43
    5.2 Immune system 48
    5.3 Liver 50
    5.4 Infertility 52
    5.5 Additional side effects 54
6. Conclusion and future research 57
7. References 59

Abstract

Synthetic derivatives of the human body's endogenous testosterone, anabolic androgenic steroids (AAS) have been used for many years, both recreationally and in a competitive sense by those looking to increase muscle size and strength. The prevalence of AAS use is not a recent phenomenon, especially in western countries where its use is on the increase.

This project aims to bring to light some of the key physiological mechanisms involved in skeletal muscle hypertrophy, including the involvement of testosterone and AAS in the recruitment of satellite cells and in increasing the number of myonuclei, leading to strength gains and an increase in muscle mass. Although this project discusses the beneficial effects of testosterone and AAS in terms of building muscle, there are also numerous side effects associated with their use, which usually occur when suprapharmacological dosages and the polypharmacy (commonly known as stacking) of AAS are taken and may lead to disorders such as cardiovascular diseases, immunodeficiency syndrome, hepatotoxicity and a range of psychological disorders amongst many others.

AAS can increase skeletal muscle mass but could also potentially cause a number of different side effects in the process, some of which will be discussed in this project.

1 Introduction

Anabolic androgenic steroids (AAS) are hormones that are produced naturally in the human body and promote a variety of effects, essential for natural physiological processes to take place. Synthetic forms of testosterone and the anabolic effects of AAS are the focus of this project and include drugs such as nandrolone (Deca – Durabolin), methendienone (Dianobol), stanozolol (Winstrol) and numerous other oral and injectable synthetic derivatives of the body's endogenous testosterone. Anabolism relates to the fact that these hormones promote cellular growth and protein synthesis and is a state in which nitrogen (reactive nitrogen species) retention occurs in lean body mass (Kuhn, 2002) which has an important role in the cell signalling pathways and mechanisms involved in skeletal muscle adaptation and muscle fibre plasticity in response to exercise and prolonged periods of muscle workload (Powers, et al, 2011). Androgenic effects causes or controls the development of male characteristics by binding to androgen receptors (AR) (Sriram, et al, 2010). ARs activated by testosterone can effect gene expression by acting as transcription factors and can cause muscle growth by stimulating protein synthesis in skeletal muscle cells (Palvimo, 2012).

Many individuals that use AAS, tend to do so for their anabolic effects; although there are side effects related to anabolism and muscle hypertrophy, many of the unwanted side effects stem from the androgenic effects of these hormones, whilst the anabolic effects of testosterone has proven to be hugely popular and widely used amongst athletes and those looking to increase fitness and athletic performance by increasing strength and muscle in both a competitive sense and for recreational use (Kadi, 2008). When testosterone was first isolated in the mid 1900s, and synthetic derivatives began to be produced, the medical community believed that it was possible to develop steroids purely for their anabolic effects and that it was possible to keep the androgenic functions of the steroids separated, thus minimizing the androgenic side effects associated with their use. However, this was later deemed to be impossible, as it was discovered that both the androgenic and anabolic effects have the same mechanism of action, but elicit their actions on different tissue types (Sterngass, 2010).

AAS use is becoming increasingly problematic with use amongst individuals using the drugs in a recreational sense on the rise. In the 21st century, many users have no competitive athletic aspirations, but aim to change their body image by becoming more muscular. In the Western World, about 2-6% of men have used AAS at some point. The reasons as to why individuals feel the need to take steroids has been the subject of plenty of scrutiny; male masculinity is strongly emphasised and rewarded in Western countries where muscularity seems to equates to masculinity, causing many men to strive to reach the levels of stardom of those they see on television, sports or posters, and they see muscularity as the way to reach those goals. Muscle dysmorphia, also known as bigorexia nervosa or reverse anorexia nervosa is a psychological issue, in which an individual is never content with their body image and no amount of muscularity or gains in strength is enough to satisfy their body image desires. Many men in Western countries see AAS as a route to reach their goals, but many are uneducated and use the drugs recklessly, which is where the many possible side effects stem from (Pope, et al, 2012).

The fact remains that ASS are highly effective and work extremely well in combination with the correct nutrition and strength training to increase muscle size and strength, but also to reduce body fat to extremely low levels (many competitive bodybuilders have a body fat percentage of 3-4% during competitions compared to average male body fat percentages of 20-30%). On average, men have more muscle and less fat than women, which is because men have higher levels of testosterone. On average, an adult woman has 8-10% more overall body fat than men (Mootz et al, 1999). The average male testis produces approximately 45mg of testosterone a week, a rate which is affected and determined by production of stimulators such as luteinizing hormone (LH) (figure1), the number of testicle LH receptors and synthesis substrate availability (Hackney, et al, 1998).

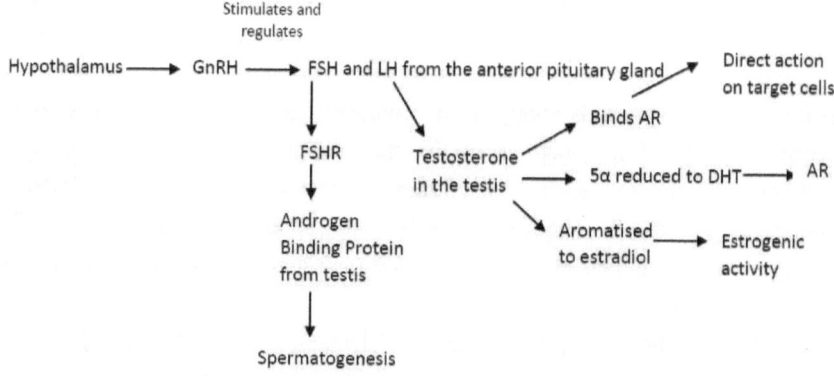

Figure 1: Male endogenous testosterone production, stimulation and regulation.

GnRH is secreted from the hypothalamus, which stimulates and regulates the production of FSH and LH from the anterior pituitary gland. FSH binds to its receptor which stimulates the production of Androgen Binding Protein in the testis. This stimulates spermatogenesis. LH can stimulate testosterone production in the testis which can either bind to an androgen receptor and directly act on target cells, can be 5α reduced to DHT which can then bind AR, or it can be aromatised to estradiol which could lead to estrogenic activity. (GnRh – Gonadotropin releasing hormone, FSH – Follicle stimulating hormone, FSHR – Follicle stimulating hormone receptor, LH – Luteinizing hormone, AR- Androgen receptor, DHT – dihydrotestosterone).

In women, testosterone production is far less, and equates to approximately 5.5mg a week, two thirds of which is secreted from ovarian and adrenal precursors (Becker, 2001). Therefore if men take additional testosterone in the form of synthetic AAS, in addition to gaining strength and muscle, men are also able to gain an exceptional amount of leanness due to fat reduction, and can gain an exaggerated amount of muscularity, particularly in the upper body, as AAS tend to selectively increase upper body mass (Pope, et al, 2012).

Any one individual who takes AAS is susceptible to being at risk from the many side effects that AAS can cause. Many people who take AAS do so for a short amount of time and then stop taking them for a few months in what is commonly known as an AAS cycle. During cycles, it is common for a user to use several AAS, including a mix of orals and injectables in what is known as stacking. AAS stacks may include anything from three to a dozen compounds, depending on the goals of the user. Those who compete in either weightlifting or bodybuilding competitions or those individuals who are looking to reach a specific target in a certain amount of time may also employ what is commonly known as pyramiding.

Pyramiding involves the gradual increase in dosage of the use of steroid stacks until competition day or until their goal such as body weight or a certain level of strength has been met, after which the dosages decrease and in most cases, the individual will stop the AAS cycle (Bomze, et al, 1991). The majority of steroid users do so for a short amount of time and are casual users, content to take AAS for a few cycles before stopping. However, approximately 30% of users go on to develop a steroid dependency, in which they take cycles over years and even decades in a continuous manner, without stopping between cycles (Pope, et al. 2012). In addition to increasing strength and muscularity, this is where many of the side effects from AAS use stem from and become much more prominent. The aim of this project is to bring to light some of the mechanisms by which AAS can increase muscle and the potential side effects this may cause.

2 Skeletal muscle hypertrophy

2.1 Muscle fibre structure

Skeletal striated muscle consists of a number of different fibres, such as type 1, 2A or 2X/B fibres. Each muscle fibre is an elongated cell that contains all of the general cellular components, such as a nucleus, ribosomes etc (figure 2).

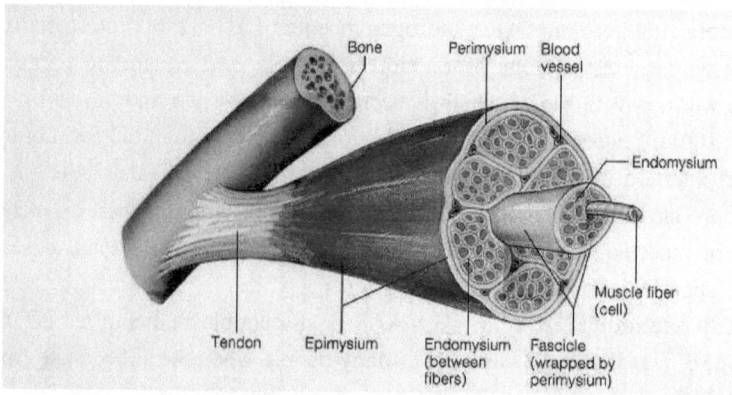

Figure 2: Connective tissue of the skeletal muscle (Image: http://www.tarleton.edu).

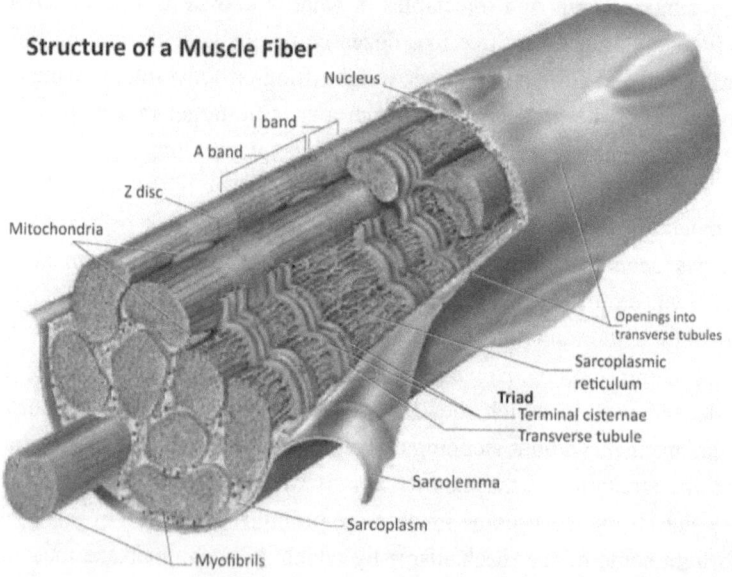

Figure 3: General structure of a skeletal muscle fibre, (Image source: http://faculty.weber.edu).

Muscle cells are composed of myofibrils which consist of the contractile element protein filaments myosin and actin. These contractile elements integrate and are linked to form sarcomeres, the basic force generating units of skeletal striated muscle (Gunther, et al, 2012). A stimulus sent to the muscle fibres by a motor neuron induces interactions between the contractile elements, regulated by actin dynamics, which causes sarcomere contraction (Awrejcewicz, et al, 2013) due the kinetics of cross-bridge cycling. These interactions can cause the muscle to either tense up, reduce in length or relax – a key process in eliciting dynamic contractions, such as those during explosive bouts of weight lifting (Telley, et al, 2007).

2.2 Muscle fibre types

The three types of muscle fibres in striated skeletal muscle are type 1 fibres, type 2 fast twitch muscle fibres, and type 2B (also known as type 2X) fibres,. Every muscle has a different composition of the three muscle fibres. For those individuals looking to increase muscle growth, it's likely that their muscle composition would consist of mainly fast twitch type 2 fibres. In previous studies investigating the composition of muscle fibres in competitive weightlifters and power lifters, it's been determined that on average, approximately 53-60% of muscle fibres in the vastus lateralis (the largest muscle of the quadriceps, recruited during squat exercises) are type 2 muscle fibres, whereas in untrained men, this percentage falls to 26-28% (figure 4). This also serves to demonstrate that muscle fibres are capable of undergoing plasticity and can change fibre type over many weight workout sessions in response to the different demands and strains placed on the muscle (Fry, et al, 2003) (Mougious, 2006). It also shows that type 2 fibres are highly responsive to muscle hypertrophy, and sensitive to intense training involving repeated muscular contractions with heavy weights, causing a large amount of force with a high tension output (Tesch, et al, 1982). There are a number of different factors that are essential for inducing muscle hypertrophy; weight training and muscle stimulation, the correct nutrition and rest and recuperation of the muscle which is essentially when hypertrophy takes place.

However, in every individual, muscle fibres respond differently to different types of weight training and nutritional intake etc. and there is a large amount of variation in the size, strength and the ability of a muscle to undergo hypertrophy from person to person (Kinirons, 2011). Hypertrophy increases muscle size by increasing the cross sectional area (CSA) of the muscle, number of ribosomes and myofibrils, thus increasing the muscle's ability to generate a greater amount of force (Horne, et al, 1990).

Figure 4: Histochemical stain for mATPase fibre type composition, taken from the vastus lateralis muscle of a non weightlifting subject (left). The image on the right is a stain taken from the same muscle type, but of a competitive powerlifter.

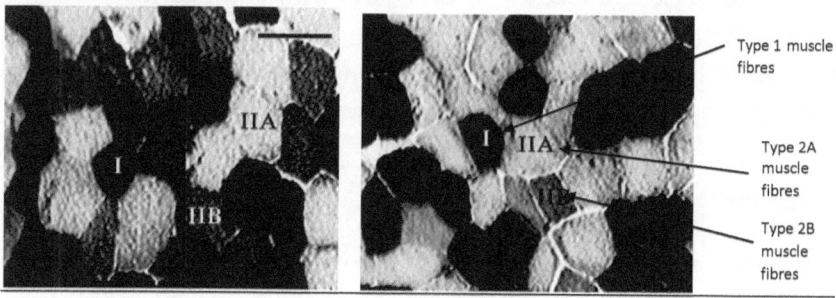

Fry, et al, 2003

2.3 Muscle hypertrophy

When subjected to stress and tension, muscle fibres may split and grow, leading to more separated and an innervated group of fibres, therefore leading to an increase in muscle fibres, mainly type 2 fibres, within motor units. Longitudinal splitting is a mechanism carried out in order to compensate for the damage exerted upon muscle fibres (Swash, et al, 1977). It is caused by the mechanical stress placed on muscle fibres, causing actin filaments to pull or displace (Kommi, 2008).

There is a positive correlation between muscle strength, the CSA of the muscle and the amount of lean body mass (Maughan, et al, 1983). ~40% total body mass in healthy humans is composed of skeletal muscle mass, and this accounts for ~30% of the resting metabolic rate. In weight training athletes, this figure is likely to be significantly higher. There are a number of metabolic pathways that provide muscles with energy during exercise (Egan, et al, 2013):

Figure 5: Metabolic pathways that fuel muscles during exercise

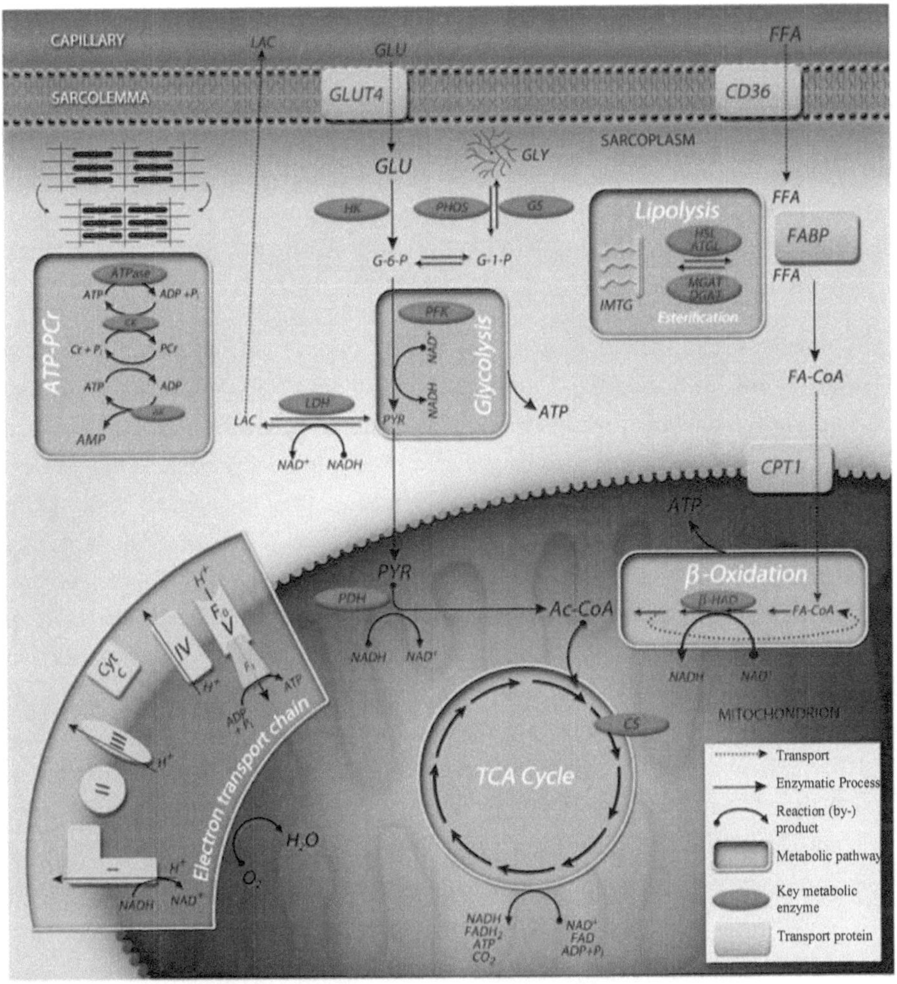

Figure 5: The three metabolic pathways that are used to fuel muscle cells during exercise. The first pathway is the ATP-phosphagen system, used during short intense workouts such as one repetition power lifting exercises. The second metabolic pathway is anaerobic glycolysis, which is used in anaerobic conditions. The third molecular pathway is the processes of carbohydrate (glycolysis) and lipid (β-oxidation) metabolism. Each of these metabolic pathways can be recruited during exercise to generate ATP. Phosphocreatine – (PCr), creatine kinase (CK), creatine (Cr), orthophosphate(Pi), adenosine diphosphate (ADP), adenylate kinase (AK),

adenosine monophosphate (AMP), muscle glycogen (GLY), glycogen phosphorylase (PHOS), circulating blood glucose (GLU), hexokinase (HK), pyruvate (PYR), lactate (LAC), lactate dehydrogenase (LDH), acetyl-CoA (Ac-CoA), tricarboxylic acid (TCA), electron transport chain (ETC) pyruvate dehydrogenase (PDH), glycogen synthase (GS), Fatty acyl translocase (FAT/CD36), fatty acid binding protein (FABPpm), carnitine palmitoyltransferase 1 (CPT1), Free fatty acids (FFAs), intramuscular triacylglycerol (IMTG), monoacylglycerol acyltransferase (MGAT), diacylglycerol acyltransferase (DGAT), Hormone sensitive Lipase (HSL),Adipose Triglyceride Lipase (ATGL) (Egan, et al, 2013).

Athletes who partake in a regular weight training programme or those who take AAS, are likely to have a greater amount of muscle mass than the average person. Those individuals who have a large amount of lean body mass, greater than the average of ~40%, will tend to have a greater proportion of muscle tissue. Muscle tissues are highly metabolically active tissues and so these individuals will tend to have a higher Basal Metabolic Rate (BMR) due to the tissues' high energy consumption (Khimji, et al, 2011). For the majority of these individuals, the aim of a weight training programme is to gain muscle whilst losing fat or keeping fat gain to an absolute minimum. Most competitive bodybuilders are able to do this, and often have a fat percentage of below 5% when they step on stage. Natural (drug free) competitors are able to do this, but those who compete in open shows (not drug tested) and take AAS, achieve remarkable muscle growth combined with an extremely low body fat percentage due to the ability of AAS to help an individual gain muscle and lose body fat (Kanayama, et al, 2009). In order for this to occur, an individual must ultimately aim to get the body into an anabolic state in which calorie and fat burning is taking place in addition to muscle building.

2.4 Additional factors affecting muscle growth; nutrition and rest

Nutrition is a key factor; those looking to increase muscle mass should ideally be consuming enough of the right calories to support a larger mass. Since muscles are highly metabolically active tissues, enough energy needs to be consumed in the form of carbohydrates and protein to aid muscle repair and growth. The type of protein consumption before and after weight training can determine the acute amplitude of muscle protein synthesis and therefore the amount of muscle gain (West, et al, 2011). Eating small and frequent meals helps to get the body into an anabolic state, as it helps to suppress the production of fat by lowering the insulin response to food. Eating meals in this way allows the energy and nutrition to be more effectively utilised by the body, without the majority of it going into fat stores, such as the formation of adipose tissue (Benardot, 2012). Sleep is also essential in maintaining a high metabolic rate, saving and replenishing energy stores (Porkka – Heiskanen, et al, 2003) and for the growth and repair of muscle.

Muscles gain energy from their fat burning capabilities. During aerobic exercise, effective combustion of fat and muscle oxygenation will build up muscle endurance. During weight training, muscles increase in their vascularity which is the result of more blood transporting more red blood cells to the muscles that are being worked (Delavier, et al, 2011).

Weight training subjects muscles to stress and tension, causing tears and strains which is followed by growth, repair and muscle reinforcement, which if combined with the appropriate nutrition, will lead to muscle hypertrophy. During a strenuous bout of weight training, muscle catabolism takes place in which the muscle becomes damaged. Rest and nutrition (protein), will allow the body to reinforce this damaged muscle structure, by synthesising actin and myosin filaments, which over time, will lead to an increase in muscle mass and build up a resistance to the catabolic effects of weight training. Muscle filaments such as actin and myosin contractile filaments are responsible for muscle contraction. It is the addition and reinforcement of these elements that leads to muscle hypertrophy. An increase in muscle fibres can also be attributed to the proliferation of satellite stem cells which have the ability to become muscle cells upon regular stress and tension to the muscle (Delavier, et al, 2011).

A high level of strength can be determined by the number of muscle fibres that an individual is able to recruit at any given time. This can be estimated as the peak force achieved during a maximal voluntary contraction (MVC) which is due to the force capacity of muscle fibres and the activity and activation of the corresponding motor neurons (Duchateau, et al, 2006).

2.5 Motor neurons and neuronal involvement in muscle contraction

Motor neurons are supplied directly to skeletal muscle fibres. Neuronal cell bodies are found in the spinal cord; axons go to the peripheral nerves by the ventral roots. Pre-synaptic nerve terminals cover the dendrites and soma and have the general synaptic machinery of a synaptic terminal, including ion channels, vesicles and a synaptic cleft. The resting potential of a motor neuron is approximately -70mV.

10mV is enough to cause a depolarisation which causes an action potential (AP) due to the increase in Na+ permeability which propagates to the nerve terminals along the axon. Motor neurons may innervate group 1A fibres, the endings of sensory axons from muscle spindles. These axons are excited by the stretching of the muscle, which then causes the muscle to undergo contraction. It is the AP in the motor neuron that initiates the muscle AP due to the release of acetylcholine (Ach) at the neuromuscular junction. This AP propagates along the muscle fibre and is dependent upon the ionic concentrations of both intracellular and extracellular Na+ and K+, membrane resistance, potential and capacitance. The AP propagates down the transverse tubule (T- tubule – 'a transverse invagination of the plasma membrane of muscle fibres', (Encyclopedia of Molecular Pharmacology, 2008) and into the interior of the muscle.

There are a large amount of L-type Ca^{2+} channels in the t-tubule membrane of muscle fibres, which undergo a slight conformational change during an AP causing charge movement. Voltage sensors are in contact with the sarcoplasmic reticulum (SR) which gates the L-type channels, and a charge movement causes the (ryanodine receptors) RyR, principally the RyR1 isoforms in skeletal muscle to open, causing SR Ca^{2+} release and an influx of Ca^{2+} into the cell. This Ca^{2+} influx causes an increase in myoplasmic free $[Ca^{2+}]$ which binds troponin C, leading to the movement of tropomyosin which causes cross bridge activation, resulting in the contractile force of the muscle (Allen, et al, 2008). ATP hydrolysis by myosin ATPase provides the initial energy source for cross bridge cycling (Egan, et al, 2013). This entire mechanism is known as the monosynaptic reflex and involves just a single chemical synapse. When group 1A fibres are stimulated, excitatory post synaptic potentials (EPSP) occur which is the result of many action potentials occurring in different pre-synaptic fibres. A large enough EPSP can cross the threshold and produce an AP, which propagates along the axons which innervate a group of muscle fibres, to the periphery which ultimately causes muscle contraction (Keynes, et al, 2011).

Once activated, skeletal muscle cells elongate and their muticellular and contractile components, such as actin and myosin filaments appear striated, hence the name, striated skeletal muscle (Keynes, et al, 2011). During strength training, a strong individual would be able to recruit and stimulate close to the maximum amount of muscle fibres in their muscle. Recruitment, stimulation and ultimately the use of these muscle fibres is attributed to the central nervous system (CNS) which generates muscle force. 'Muscles receive a neural activation signal (determined by the sum of the spiking activities of motor neurons, also known as the neural drive of the muscle) from a pool of innervating motor neurons'. The neural drive to the muscle may involve a small number of motor neurons, but will still apply to a large proportion of the working force of the muscle (Farina, et al, 2014). Studies have shown that neuromuscular electrical stimulation (NMES) performed under isometric conditions, conditions in which no visible muscle movement occurs (Chu, et al, 2013), with the application of an electrical stimuli has an impact on muscle performance and can improve an individual's isometric maximal voluntary strength (Gondin, et al, 2011).

From the cerebral level, information is sent to the neurons in the spinal cord and is then relayed to the muscle fibres due to the motor output of motor neurons, giving the muscle a command to contract. The contraction of each individual set of muscle fibres, Type 1, 2A, 2X, is determined by an individual motor neuron, as demonstrated by figure 6. Therefore the contractile force of a muscle is dependent on the amount of motor neuron activity (Duchateau, et al, 2006), which can be stimulated with the use of heavy weight training, which recruits more units. Therefore it would be a reasonable to suggest that because AAS causes muscle hypertrophy, this would lead to the recruitment of more motor neurons.

However, recent studies have actually shown that the use of AAS can cause motor neuron death or a deterioration in motor neuron function, which is contradictory to the previous statement (Aggarwal et al, 2014).

Figure 6: Motor neuron innervations of skeletal muscle

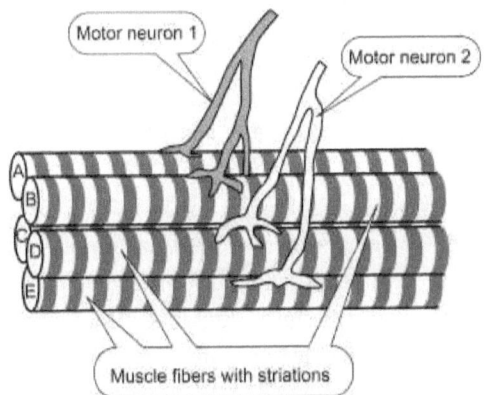

Figure 6: Diagram showing skeletal striated muscle innervated by motor neurons e.g. motor neuron 1 innervates the fibres A, B and C and motor neuron 2 innervates fibres D and E. If only motor neuron 1 fires an action potential, then only fibres A, B and C contribute to the muscle contraction. If only motor neuron 2 fires an action potential,

only D and E muscle fibres will contract. If both motor neurons fire action potentials, muscle contraction would occur at greater intensity due to a greater number of muscle fibres recruited to lift the weight (Feher, 2012).

Figure 7: A) Mice performed a hanging wire test and B) a rotarod test.

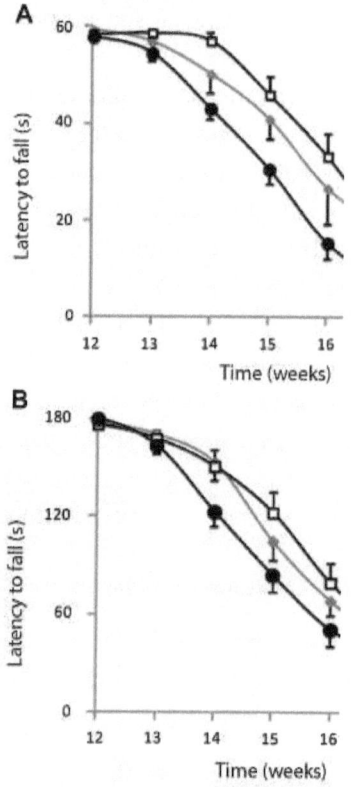

Figure 7: A) Mice performed a hanging wire test and B) a rotarod test. Over time, the mice that had been administered with the AAS nandrolone decanoate (dark black line) showed a significant decline in motor performance, in comparison to the control group (grey) and the castrated male mice (thin black line) when performing the two tasks. (Aggarwal et al, 2014). This demonstrates that steroid use may not lead to the recruitment of more motor neurones but may lead to neuronal decline.

Motor activity sent to the muscle fibres can vary in frequency. If motor neurons send electrical impulses to muscles at a low frequency, muscle contraction will not reach its maximal contractile force. If there is a high frequency of electrical impulses sent to the stimulated muscle fibers by motor neurons, muscle contraction will be at a greater intensity, which can be developed using a heavy weight training programme (Delavier. F, et al, 2011) (Williams, 2001).

3 Testosterone

3.1 Testosterone and its synthetic derivatives

AAS and synthetic derivatives of testosterone that are currently available include: androgens, such as testosterone and androstenedione or their precursors that are endogenously produced; and exogenous synthetic derivatives of testosterone that have been produced to have different levels of metabolism and receptor binding affinities.

Nandrolone, for example, has been found to have a greater binding affinity to the AR in skeletal muscle than methenolone, which in turn has a greater binding affinity than testosterone (Saartok, et al, 1984).

AAS are derived from cholesterol and are a family of lipophilic hormones, including testosterone and its synthetic derivatives, which when used in combination with weight training, exercise and nutrition, can reduce body fat, increase muscle mass and strength beyond the body's natural capabilities. Therefore AAS are becoming increasingly more popular amongst those looking to gain more strength, muscle and become leaner (Kanayama, et al, 2009).

Testosterone promotes protein synthesis and cellular growth. It is produced naturally in the human body, produced by Leydig cells and secreted by the testis in males (~95% of the body's testosterone, Kicman, 2008) and a small amount is secreted by the ovaries in females. Some testosterone is also secreted by the adrenal cortex, however in terms of building muscle, the testosterone secreted by the testis is of physiological significance (Gard, 2001). On average, the testis produces about 20 times the amount of testosterone that is produced by the ovaries on a daily bases, and so the anabolic and androgenic effects of the hormones are more pronounced in males (Southren et al, 1965, 1967). The testosterone production rates of the testis, on average falls between 3 - 17mg daily (Korenman, et al, 1963), and is influenced by the individuals lifestyle choices, such as nutrition, sleep and exercise. The body's natural testosterone production increases directly following a heavy bout of exercise, such as weight resistance training (Vingren, et al, 2010). Studies have also been conducted on whether a person's diet, specifically protein - carbohydrate ratio regulates testosterone production; Anderson et al (1987), discovered that the testosterone concentrations in men increased after 10 days of a high carbohydrate diet rather than a high protein diet (Anderson, et al, 1987). Efficient sleeping patterns have shown to increase the amount of bioavailable circulating testosterone and sleep disturbances decreases overnight plasma bioavailable testosterone (figure 7) (Schiavi, et al, 1992).

Figure 7: the effects of sleep on testosterone

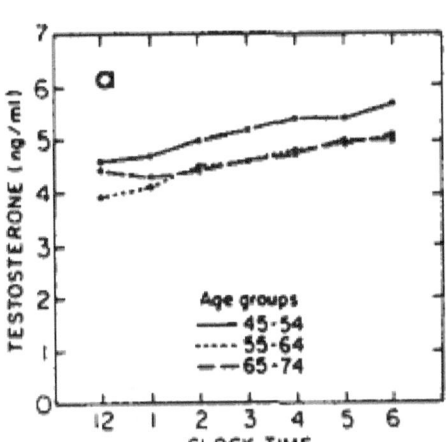

Figure 7: The effects of sleep at certain times of the night on testosterone levels of the men of three different age groups. The 45-54 age group (the top line) had a greater increase in testosterone during sleep (Schiavi, et al, 1992).

Since testosterone is known to promote muscle hypertrophy, especially when combined with strength training (Bhasin et al, 1996), those looking to increasing muscle size and strength often adhere to the principals of strength training, an efficient amount of sleep and rest, and the correct nutritional requirements.

There are many different synthetic derivatives of natural testosterone (figure 8):

Figure 8: Chemical structure of testosterone and synthetic derivatives

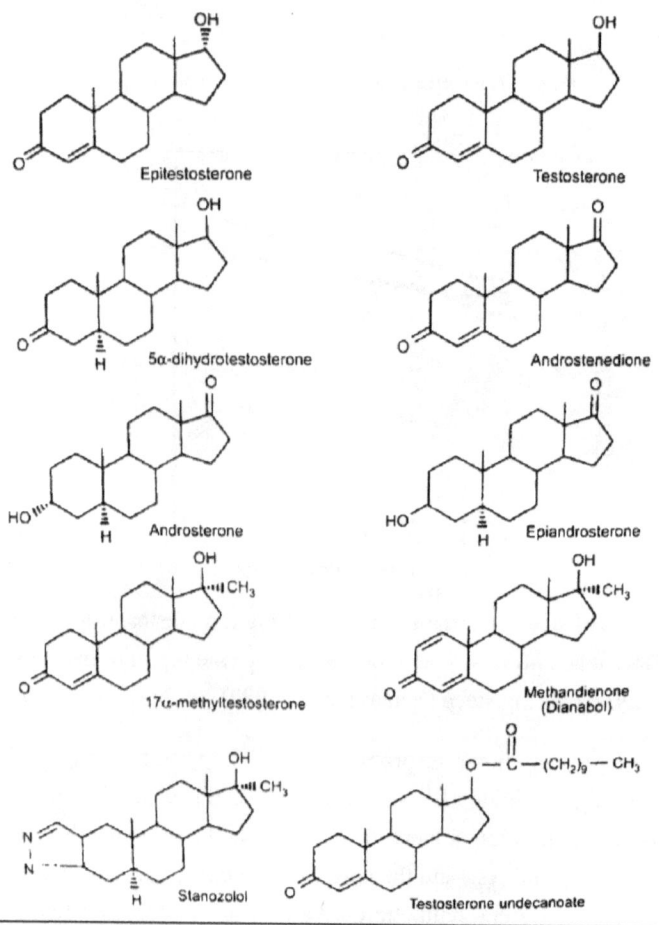

Figure 8: The chemical structure of testosterone and some of its natural and synthetic derivatives (Mottram et al, 2000).

Administering pure testosterone, either through oral administration or by intramuscular injections is not considered as effective for those who wish to gain muscle, due its rapid absorption rate to the portal blood system and metabolization in the liver. Injecting pure testosterone intramuscularly should also act to minimize the first pass metabolism of the drug. Therefore synthetic derivatives of testosterone, as an ester or AAS are now widely used (Schanzer, 1996).

3.2 Testosterone mechanisms of action

Testosterone and male sex hormones in general are known as androgens. They can act on both reproductive and non-reproductive tissues and in male development, they have a crucial role in the development and formation of various male characteristics. In the uterus for example, they play a role in the development of male genitalia, during puberty they are crucial for the development of secondary male characteristics such as the growth of facial hair and long bone growth and can lead to mood changes and psychological changes during adolescence, such as aggressiveness and an increases libido. They can also exert their effects in combination with other hormones, such as the follicle stimulating hormone (FSH) which is secreted from the anterior pituitary gland and is crucial in the production of sperm. Androgens are also crucial for muscle development to occur, which has been recognised and has lead to synthetic versions being produced (Gard, 2001). It is these muscle building anabolic properties rather than the androgenic and masculinisation effects that people hope to gain by using AAS, in order to improve body image, size and physical appearance (Ranjan, et al, 2014).

Both endogenous testosterone and AAS have anabolic and androgenic effects, but the majority of people that use AAS do so for the anabolic effects that they have on muscle. Depending on the individual user and aside from the anabolic effects, many of the unwanted side effects stem from the androgenic activity of the drugs (table 7). Such effects are more prominent in female AAS users due to the obvious anatomical sex differences between male and female users. Women have lower baseline levels of testosterone than men, which means that when AAS are administered, they are likely to be more sensitive and more prone to side effects (Ip., et al, 2010). Most AAS manufacturers, including those licensed to produce the drugs and underground laboratories, aim to produce AAS that have a greater anabolic activity, however each individual drug has varying degrees of balance between anabolic and androgenic activity, and as of yet, there has not been any AAS produced that has purely anabolic activity (Kuhn, 2002).

Figure 9: The mechanism of action of testosterone

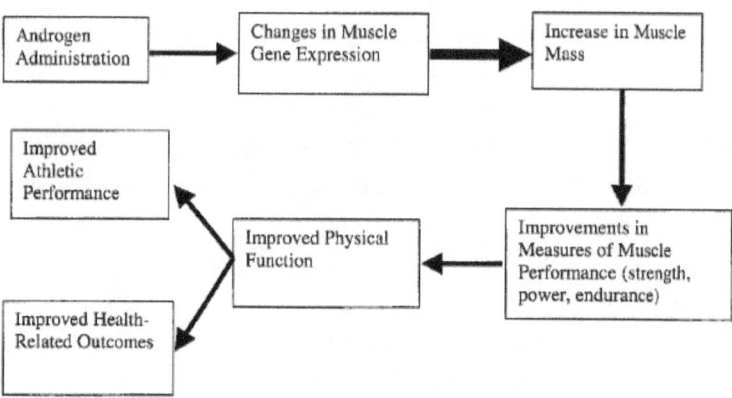

Figure 9: The mechanism of action of testosterone. The effects of steroid use is still under research, so the thickness of the arrows correlates to the levels of certainty; thicker the arrow the greater the certainty (Bhasin et al, 2001).

3.3 Testosterone effects on skeletal muscle fibres

Testosterone increases the muscle hypertrophy of skeletal muscle, primarily type 1 and type 2 muscle fibres (Kadi, 2008). The majority of AAS users and weightlifters or strength trainers in general have an abundance of type 2 muscle fibres, including type 2X and type 2A fibres, and have a relatively fewer amount of type 1 muscle fibres. Type 1 fibres are engaged primarily during endurance exercises, and can contract over a relatively long period of time without depleting or getting tired. Also known as slow type 1 muscle fibres, they have a large amount of mitochondria and mitochondrial enzymes, produce ATP and are known as oxidative, as they rely heavily on oxidative processes in order to drive the production of ATP. These fibres have a high triacylglycerol and myoglobin concentration. Myoglobin is the primary oxygen carrying pigment of muscle tissues and therefore a large amount of myoglobin is essential in an individual's performance level during endurance exercises (Ordway, et al, 2004; and Mougios, 2006).

Type 2X/B fibres are glycolytic and can be considered to be opposite to type 1 fibres, in that they have low triacylglycerol and myoglobin concentrations, in comparison to type 1 fibres there are a relatively small amount of capillaries surrounding the muscle and they have a low mitochondrial density. Type 2X/B fibres have a high concentration of lactate dehydrogenase and produce a large concentration of lactate during muscle contraction which can in tern disrupt the functioning of other metabolites in the glycolyitic pathway, increase acidity in muscle cells and can cause muscles to cramp. ATP is regenerated due to anaerobic processes and the muscle relies on glycogen as a source of energy when the muscle is recruited during fast and explosive bouts of exercise, such as weightlifting or power lifting (Mougious, 2006).

Type 2A fibres are similar to type 2X/B fibres, although they exhibit many of the properties of type 1 fibres. These fibres are known as oxidative-glycolytic fibres and can undergo fast contractions over a relatively long amount of time, in comparison to type 2X/B fibres, without getting tired (Mougious, 2006).

Muscles contain all three fibre types. Genetics plays a role in determining what types of muscle fibres a certain individual may have, but over time, if a heavy workload is placed on a particular fibre and the demand for that fibre increases, muscle plasticity does allow fibres to change into a different type due to the functional demands and requirements of the body. Those individuals looking to increase muscle size and strength and perform better in their workouts, look to increase primarily type 2 muscle fibre types (Mougious, 2006).

The main effect of testosterone and AAS is that they increase fat free muscle mass in skeletal muscle. There are numerous studies that have been conducted that show testosterone and its synthetic derivatives increase fat free mass in skeletal muscle (figure 10).

Figure 10: The results of six studies showing the effects of testosterone and AAS on skeletal muscle (Bhasin et al, 2001).

Study	Age (years)	Testosterone regimen	Change in fat-free mass
Bhasin et al. (1997)	19–47	Testosterone enanthate 100 mg weekly for 10 weeks	5.0 ± 0.7 kg ($9.9 \pm 1.4\%$) increase by underwater weight and D_2O
Katznelson et al. (1996)	22–69	Testosterone enanthate or cypionate 100 mg weekly for 18 months	$7 \pm 2\%$ increase by bioelectrical impedance
Brodsky et al. (1996)	33–57	Testosterone cypionate 3 mg/kg every 2 weeks for 6 months	15% increase by DXA scan
Wang et al. (1996)	19–60	Sublingual testosterone 5 mg three times a day for 6 months	0.9 kg (2%) increase by DXA scan
Snyder et al. (2000)	22–78	Transdermal testosterone patch for 12–36 months	3.1 ± 3.3 kg increase by DXA scan
Wang et al. (2000)	19–68	Testosterone gel (50–100 mg/day) × 180 days	2.7 ± 0.3 kg increase by DXA scan

D_2O, deuterium water.

3.4 The action of testosterone on myonuclei and satellite cells

Rather than increasing the relative proportion of these muscle fibres, testosterone increases the number of myonuclei and satellite cells (Choong, et al, 2008) as determined in a study conducted by Sinha-Hikim et al (2003) (figure 17).

Myonuclei are the muscle fibre nuclei and are an important determinant of DNA for gene transcription (Shenkman, et al, 2010). Transcription factors such as TEF-1 are important for the regulation of gene expression in skeletal muscle and rely on factors such as the muscle specific cofactor Vestigial- like-2 (Vgl-2) which has been shown to be important in the promotion of skeletal muscle differentiation and the activation of muscle specific promoters (Chen, et al, 2004). Skeletal muscle satellite cells are inactive mononucleated myogenic cells that lie next to differentiated muscle fibres. They can proliferate from a reserve pool of cells and by doing so can form regenerated muscle and additional satellite cells. These satellite cell muscle precursor cells are essential in muscle hypertrophy. They also have medical uses, including being used in muscle regeneration after skeletal muscle damage and can also be used as a vector in possible gene therapy (Morgan, et al, 2003). Satellite cells have the ability to proliferate and differentiate into myotubes which can fuse together to reinforce muscle fibres. It may be the case that during this fusion process, the centrally located myonuclei may become trapped and be required for myotube fusion with the parent muscle fibre, resulting in a larger muscle fibre to be formed (Kadi, 2008).

3.5 Androgen Receptor (AR) expression

Expression of androgen receptors (AR) occurs in satellite cells and in the myonuclei of some skeletal muscle fibre types (Choong, et al, 2008). This was confirmed using staining techniques as seen in figure 11, by Sinha-H, et al., leading to the conclusion that satellite cells are the main site of AR expression in skeletal muscle, supporting the fact that androgens can lead to muscle hypertrophy due to their expression on several cell types in order to regulate the differentiation of mesenchymal precursor cells in the skeletal muscle (Sinha-H, et al, 2004).

Figure 11: Histochemical stain of human skeletal muscle cultures

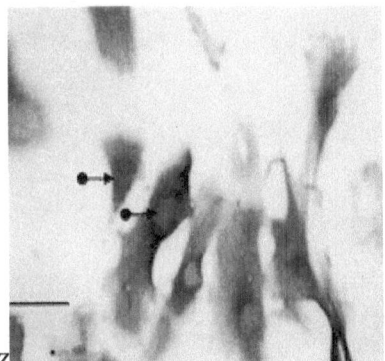

Figure 11: Histochemical stain of human skeletal muscle cultures showing that the AR were enriched in satellite cells. The cells show AR expression localisation (indicated by arrows) in the cytoplasm and nucleus after they had been incubated with 10nM testosterone. (Sinha-H et al, 2004).

The AR is activated and regulated upon the binding of the androgen, which undergoes a series of conformational changes, specifically in the ligand binding domain (LBD) affecting AR-protein and AR-DNA interactions, and associates with its beta-catenin co-activator.

These conformational changes can affect AR interactions with other proteins and can cause the dissociation of heat shock proteins. AR-protein interactions are stimulated upon androgen binding, but are also necessary to stimulate the AR transcriptional activity. These proteins are known as coregulators that can enhance or reduce the transactivation of target genes and can facilitate AR conformation, nuclear localisation and DNA binding and interactions with transcriptional machinery. Based on their chemical structure, affinity for the AR, activating or inhibiting effect on the transcription of AR target genes, AR ligands can either be classified as either steroidal or non-steroidal. (Choong.K, et al, 2008; Fragkaki. A.G, et al, 2009). Figure 12 gives an overview of the AR structure, function and some of the key interactions involved.

Figure 12: Testicular synthesis occurs and testosterone is transported to skeletal muscle tissue.

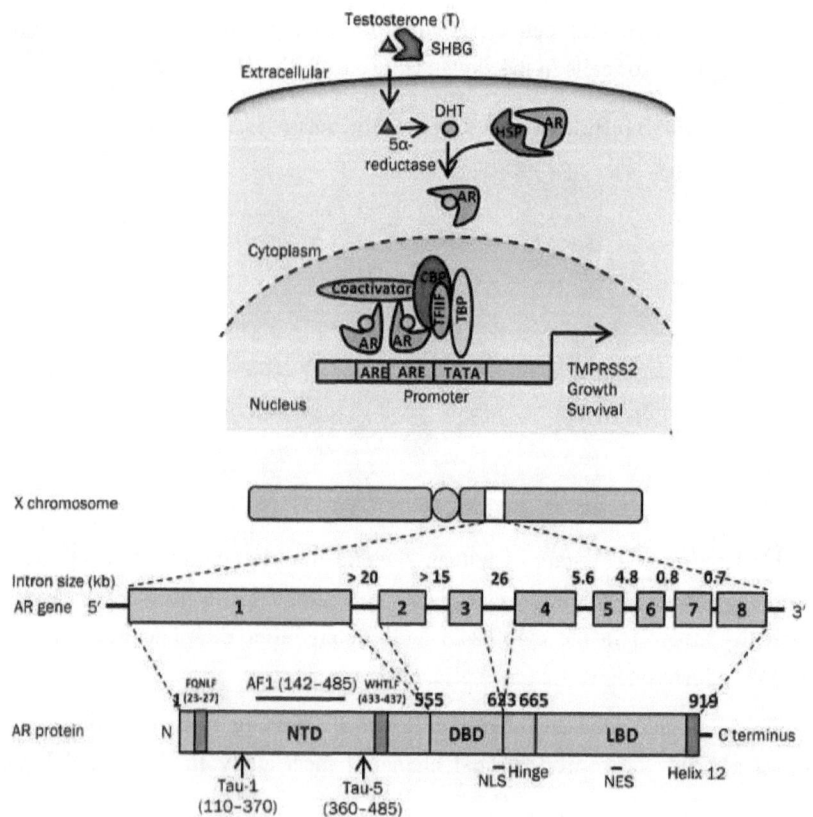

Figure 12: Testicular synthesis occurs and testosterone is transported to skeletal muscle tissue. It's transported in the blood by sex hormone binding globulin and is converted to dihydrotestosterone (DHT) by 5-α-reductase. This then binds to the binding pocket of the ligand which causes heat shock proteins to dissociate from the AR. This causes the translocation of the AR into the skeletal muscle cell nucleus, where it undergoes dimerisation and binds the androgen response element in the promotor region of the target gene. Here, the AR recruits transcriptional machinery and other coregulators. The AR gene contains eight exons and introns of different lengths (Tan et al, 2015).

3.6 Hormonal signalling mechanisms and endogenous testosterone production

Gonadotropin releasing hormone (GnRH) is secreted by the hypothalamus, as determined by Hall et al (Hall, et al, 1992). In turn, this both stimulates and regulates the secretion of luteinizing hormone (LH) from the gonadotroph cells in the anterior pituitary gland, endocrine cells that produce gonadotropins, and follicle-stimulating hormone (FSH), also synthesised and secreted by the gonadotrophs in the anterior pituitary gland. In men, FSH stimulates and enhances the production of androgen binding protein (ABP), a beta globulin glycoprotein that is produced by the sertoli cells in the testis by way of binding to the transmembrane receptor, follicle-stimulating hormone receptor (FSHR) on their basolateral membranes (Boulpaep, et al, 2005). This in turn, stimulates spermatogenesis (see figure 1). LH, a heterodimeric glycoprotein similar in structure to FSH, stimulates the Leydig cells of the testis to produce testosterone. LH also regulates the amount of testosterone production which regulates the expression of 17-β hydroxysteroid dehydrogenase (17β-HSD) (Payne, et al, 1995). 17β-HSD are alcohol oxidoreductases and have a vital role in controlling the potency of steroid hormones such as testosterone by catalyzing oxidation or reduction at position 17 in their chemical structure (Adamski, et al, 2001) and mainly the catalysation of the reduction of androstenedione to testosterone (Mohan, 2008) by type 3 and 517β-HSD in the testis and peripheral tissues . There are various types of 17β-HSD, which have tissue specific expression properties and substrate specificity which works to provide periphery cells with the mechanisms required in order to control the concentrations of intracellular androgens. 17β-HSD have a medicinal use and are used as a form of hormonal control known as intracrinology (Labrie, et al, 1997). Once LH has stimulated the production of testosterone in the testis, testosterone can act on target cells, either directly, or it can be 5α reduced by the action of 5α- reductase enzymes. These can convert testosterone to dihydrotestosterone (DHT). AAS may be produced to target this conversion of testosterone – figure 13. DHT can also act on the AR and is more active than testosterone (Gradisnik, et al, 2009), or testosterone can be aromatised to estradiol by way of the aromatase enzyme complex.

Figure 13: Sites that may be the targets for modification for AAS production for 5 α DHT to the AR

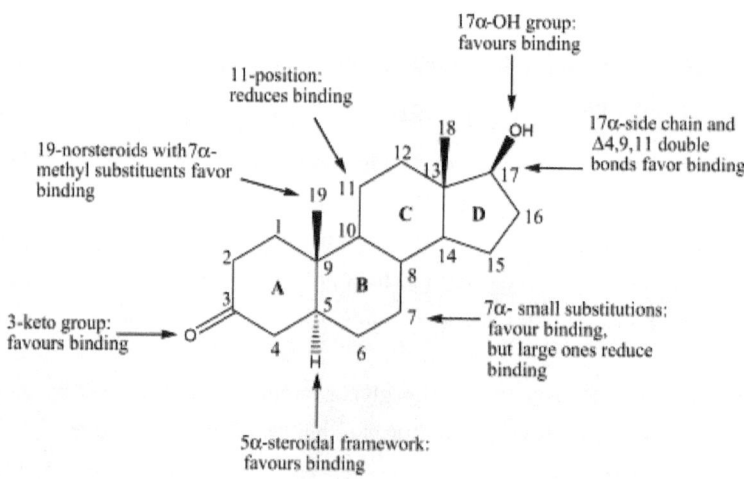

Fragakaki et al, 2009

If testosterone is converted to estradiol, it exerts its estrogenic activity, which is an undesirable activity for anabolic drugs to have; the action of the 5α-reductase enzyme reduces the anabolic:androgenic ratio of activity and the aromatisation of testosterone to estradiol results in possible feminising side effects, such as the development of female breast tissue which is a common occurrence in adolescent males undergoing hormonal changes (Kuhn, 2002). Testosterone binds to androgen receptors on the target cells, as does DHT if testosterone undergoes conversion, although DHT has a greater potency and greater androgenic effect than testosterone due to its high affinity for the androgen receptor. There is a small amount of free testosterone in the blood stream (1-2%), with most testosterone circulating in bound form, bound to the sex hormone binding globulin protein (SHBG), although this form of testosterone is not generally available or utilised by tissues.

However, SHBG may exert its action by binding non-receptor cell proteins at the cell membrane. When SHBG binds steroids, it may reach its receptor and act as a docking station, allowing testosterone and other steroids to exert their actions on the cell by transducing signals via G-Proteins, activating adenylate cyclase and subsequently cAMP which may affect the transcriptional activity of intracellular receptors such as the AR for testosterone and other AAS hormones (Li, et al, 2009). Testosterone and its various synthetic derivatives generally exert their muscle building effects by binding to the AR (nuclear receptor) in target tissues, by increasing the size of muscle cells (Bagatell, et al, 1996). Once bound to the AR, an AR complex is formed in the cell nucleus (figures 11 and 12). The AR consists of different domains, such as the N-terminal region, a DNA binding domain (DBD), and a C-terminal ligand binding domain (LBD). It also consists of hormone dependent transcriptional activation domains. After the formation of the receptor complex, translocation occurs into the nucleus, and the complex binds to hormone response elements (HRE). This binding is due to the steroid receptor specific zinc fingers in the DBD. The AR complex translocates to specific binding sites on chromatin, where it binds, resulting in gene transcription, synthesising mRNA from the DNA in the cell's nucleus which initiates protein synthesis and leads to muscle hypertrophy (Gradisnik, et al, 2009).

3.7 Possible non-genomic mode of action of testosterone

Testosterone may also have a non-genomic mode of action that may be AR-independent, although not much is known about this possible pathway. Testosterone may bind directly with a GPCR (G-Protein Coupled Receptor) of the MAP K (mitogen activated protein kinase) family at the plasma membrane of myoblasts. Elevated levels of Ca^{2+} can trigger the activation and phosphorylation of ERK1/2 (extracellular signal- regulated kinases) pathways (Schmitt, et al, 2004), as shown in figure 14.

Figure 14: The effects and time course of phosphorylation of testosterone and the steroid nandrolone on ERK1 and 2. Results were normalised and expressed as a percentage

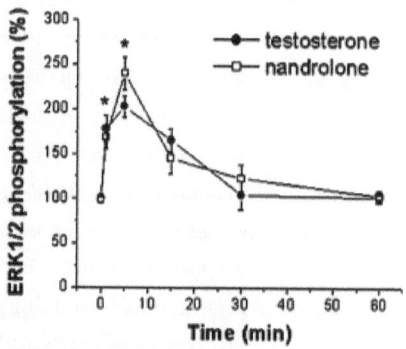

Estrada et al, 2003

The action of testosterone on the GPCR may be exerted due to an increase in $[Ca^{2+}]I$. Studies have shown that testosterone can increase Ca^{2+} signals, as can AAS such as nandrolone. This suggests that Ca^{2+} has a key role in AAS-AR interactions and that this possible non-genomic pathway is androgen specific. Although relatively unclear, figure 15 shows that there was an increase in calcium that reached 50 fold the basal value with administration of 100nM nandrolone, whereas the estrogens did not have the same effect. (Estrada et al, 2003).

Figure 15: The effects of different steroids on calcium signals in skeletal muscle cell myotubes. Relative fluorescence relates to the expression of intracellular calcium over the time course of steroid induced calcium release.

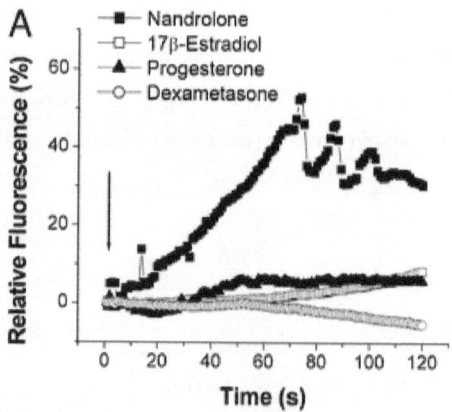

Estrada et al, 2003

This increase in Ca^{2+} would stimulate the ERK pathway signalling cascade. Upon GPCR activation by the binding of testosterone and the exchange of GDP (guanosine diphosphate) for a GTP (guanosine triphosphate), recruitment of Grb2 (growth factor receptor bound protein, adaptor protein) and SOS (a guanine nucleotide exchange factor) would take place. In turn, SOS activates Ras, which activates Raf due to phosphorylation which occurs at several sites at the plasma membrane. MEK1/2 (mitogen-activated protein kinase kinase) enzyme then undergoes phosphorylation at two serine residues which then phosphorylates ERK1/2 on threonine and tyrosine residues. ERK1/2 undergoes phosphorylation and becomes activated which then phosphorylates RSK. RSK and ERK then translocate to the cell nucleus where they have an effect on many different transcription factors, binding to DNA and other cellular proteins to control the cells' genetic machinery which ultimately controls and effects protein synthesis leading to muscle hypertrophy by changing the amount of cell proliferation. The actions of ERK, MEK and Raf are key components of the pathway, and any alterations or inactivation at these steps can stop protein synthesis and create a negative feedback loop (Mebratu, et al, 2009). This AR-independent non-genomic mode of action of testosterone provides a possible mechanism by which testosterone can bind a GPCR directly and stimulate a signalling cascade which regulates the cells' transcriptional machinery, leading to protein synthesis and muscle hypertrophy. However the extent to which this mechanism occurs in skeletal muscle cells and myonuclei still requires further research (Kadi, 2008).

4 Anabolic Androgenic Steroids (AAS)

4.1 Exogenous testosterone administration

AAS ultimately have very similar mechanisms of action as the body's endogenous testosterone. They provide the user with enhanced effects that baseline levels of endogenous testosterone has on muscle. Modifications are made in order to influence the pharmacokinetics of the drug (figures 13 and 24), its bioavailability, and to positively skew the balance of activity towards it eliciting anabolic effects (Kuhn, 2002).

It has been shown by Spiering et al (2009) that exogenous testosterone administration in the form of AAS potentiates gains in muscle strength and mass in combination with weight resistance training, that GnRH inhibits endogenous testosterone release and thus inhibits gains in muscle mass and that AR antagonists impair muscle growth. The effects of GnRH as an endogenous testosterone inhibitor and the effects of testosterone on muscle fibre hypertrophy was demonstrated in a study conducted by Sinha-Hikim et al (2003), in which healthy men were treated with injections of long lasting GnRH every month for a period of 20 weeks. GnRH suppressed endogenous testosterone secretion. Data from the study also showed the effects of exogenous testosterone on type 1 and 2 muscle fibre hypertrophy and could therefore go some way in aiding the understanding of the effects of AAS on muscle fibres. Exogenous testosterone was administered to the subjects in the form of testosterone enanthate (TE) and given to different patient in varying doses of 25-600mg for the 20 weeks. Magnetic resonance imaging was used to measure muscle volume, which increased in proportion to the increasing doses of TE. This was determined to be due to the hypertrophy of both type 1 and type 2 muscle fibres which increased, again in proportion to the increasing doses of TE (Bhasin, et al, 2003; Sinha-Hikim et al, 2003). Therefore androgen signalling is vital in the understanding of muscle hypertrophy (Spiering, et al, 2009).

Figure 16: Muscle biopsies taken before and after a period of AAS use

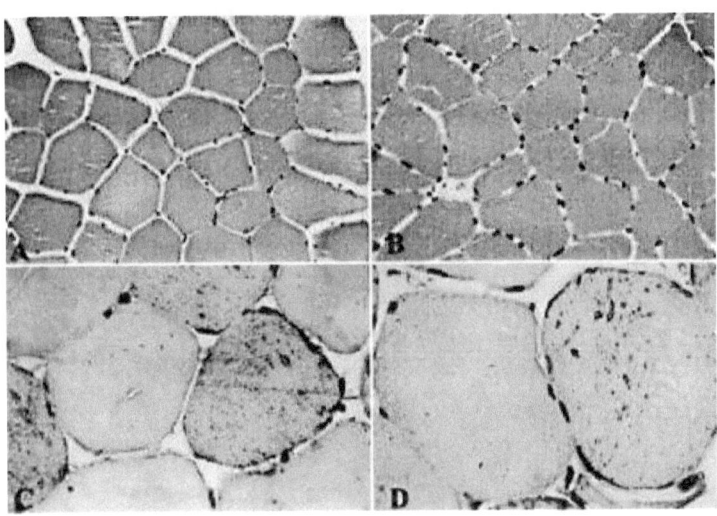

Figure 16: Cross sections of muscle biopsies taken before, A and C, and after the 20 week study, B and D. In this case, the man was treated with a weekly GnRH agonist and 600mg of TE (Bhasin, et al, 2003) (Sinha-Hikim et al, 2003).

Figure 17: Effect of testosterone on Myonuclei and Satellite cell number

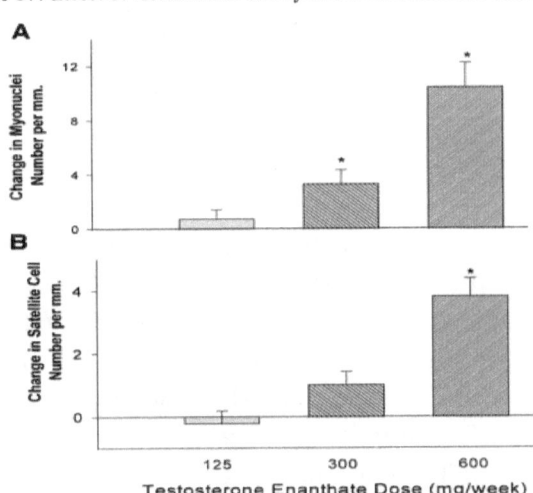

Figure 17: The effect of different doses of TE on myonuclear number (*A*) and absolute satellite cell number (*B*). Increasing the dosage increased both satellite cell and myonuclear numbers (Sinha-Hikim et al, 2003).

Weight resistance training has an impact on androgen signalling (figure 18). It increases endogenous testosterone levels which in turn increases the amount of testosterone-AR binding which potentiates gains in strength and muscle mass.

Figure 18: Mechanisms of action of testosterone on the AR

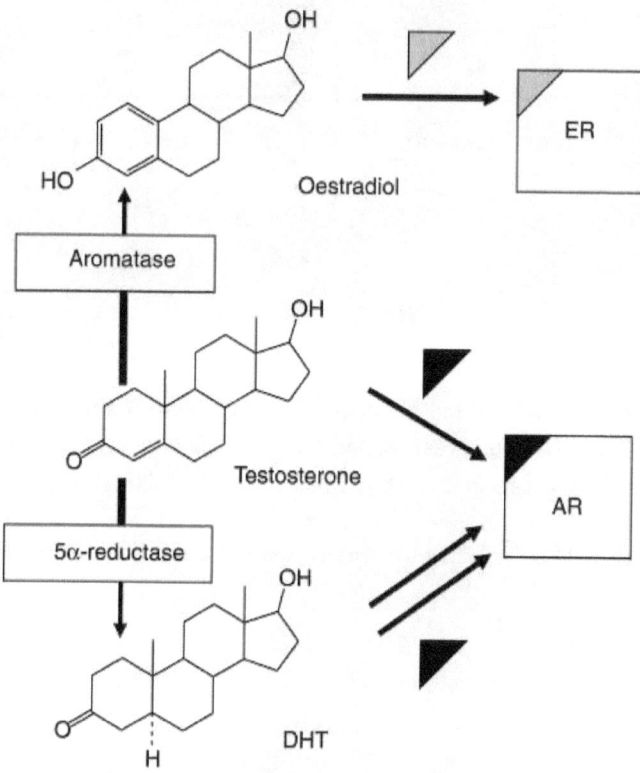

Figure 18: Testosterone can bind directly to the AR or mediated by metabolism in target tissues due to the action of intracellular enzymes such as 5α-reductase. 5α-reductase converts testosterone in target tissues to 5α- dihydrotestosterone (DHT) which has a greater binding affinity to the AR. Testosterone can also be converted by the aromatase enzyme to oestradiol which has binding affinity for the oestrogen receptor (ER). Depending on the target tissue's metabolic and enzymatic activity, both testosterone and DHT may be converted to weaker androgens (Kicman, 2008).

During weight training, up-regulation of muscle AR content due to enhanced transcription of mRNA takes place when muscle contractions occur. This increase in

AR content causes the muscle to become sensitised to the increased testosterone levels which can lead to hypertrophy in those specific muscle fibres. Weight training may increase the concentration of endogenous testosterone and therefore weight training may be capable of regulating muscle AR content. Chronic exogenous testosterone administration in the form of AAS enhances muscle AR adaptations, leading to muscle hypertrophy (Spiering, et al, 2009).

AAS are manufactured by modifying the chemical ring structure of testosterone, which may include alkylation at the 17α-position. This is ultimately carried out in order to make testosterone derivatives that are more anabolic. Esterification of the 17β- hydroxyl group carried out by carboxylic acids may also take place which enables the derivative to have lipophilic properties, giving it the ability to retain and dissolve in fats. This property can make this synthetic derivative highly popular amongst those looking to gain muscle and lose fat and it also works by prolonging and increasing the action of steroid activity. In addition to these modifications, these synthetic derivatives can also bind to AR with varying affinities (Fragkaki, et al, 2009).

AR gene expression modulation, testosterone binding affinity to AR, AR protein stability, nuclear translocation and transactivation are all important factors in the regulation of AR activity. Crytallography has brought to light information regarding the action of testosterone on AR. Testosterone–AR binding involves hydrophobic interactions in which the steroid skeleton interacts with the LBD. The hydrophobic residues of the LBD side chains are important for the stabilisation of testosterone in the LBD but are also responsible for the high specificity and selectivity of the AR. Other factors that affect testosterone binding affinity to the AR includes a steric change at the 7α position which enhances affinity by 10- fold (Fragkaki, et al, 2009).

4.2 Anabolic effects of testosterone and AAS synthetic derivatives

Figure 19: Mechanisms of action of testosterone and AAS leading to muscle hypertrophy

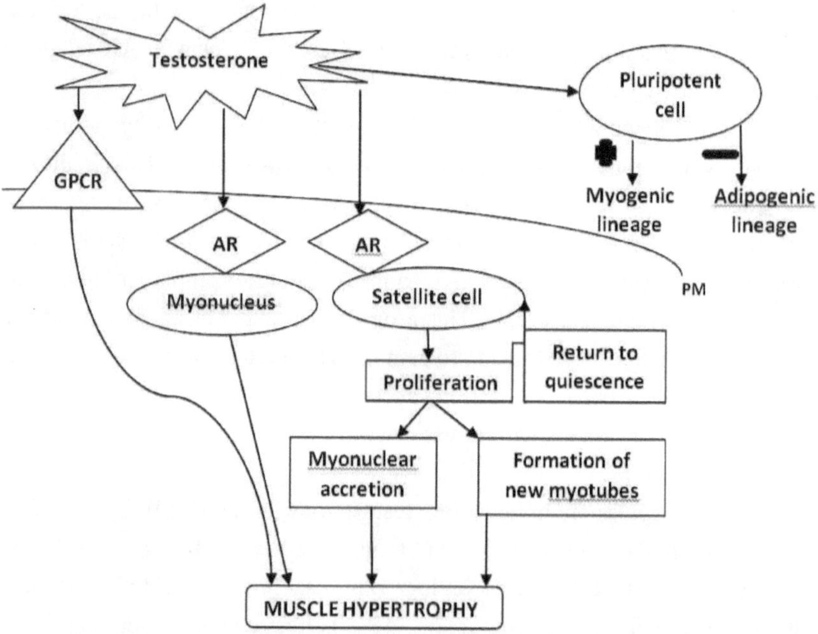

MUSCLE HYPERTROPHY

Figure 19: The action of testosterone on skeletal muscle to increase muscle hypertrophy via several myogenic pathways (adapted from: Kadi., 2008).

AAS exert their effects by effectively mimicking the mechanism of action of testosterone. Testosterone and AAS mainly increase muscle hypertrophy and exert their anabolic effects due to their interactions with AR, but also, potentially through a number of different mechanisms. Weight training increases AR mRNA, the stability of which is key in the regulation of AR expression (Yeap, et al, 2004), and the binding potential of mitogens such as testosterone. If weight training induces an increase in AR number and availability, this means that more binding sites are available; combine this with the fact that weight training increases endogenous testosterone levels (West, et al, 2011) and with the administration of AAS, and there could potentially be an increase in testosterone-AR binding which may induce the several pathways demonstrated in figures 20 and 19, leading to muscle hypertrophy (Kuhn, 2002).

Figure 20: Weight training leads to and increased amount of AR-testosterone binding

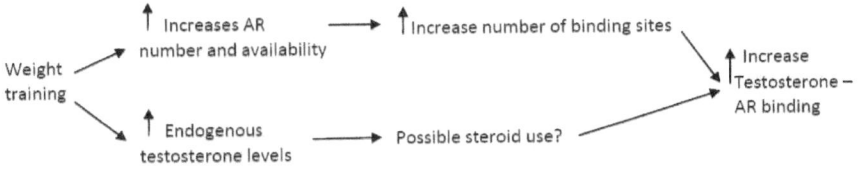

Figure 20: Weight training leads to and increased amount of AR-testosterone binding which would induce the myogenic pathways depicted in figure 19.

Pluripotent cells are essentially stem cells that have the ability to be able to divide into specialised cells such as muscle cells (Dodet, et al, 2001). They can assume either a myogenic or adipogenic lineage. Myogenesis is effectively muscle growth and regeneration, particularly during embryonic development (Pavlath, 2011), although it has been widely speculated in the field of stem cell research that adult stem cells and embryonic stem cells do have very similar functional capabilities (Wang, et al, 2010). Adipogenesis is the process by which pre-adipocytes derived from embryonic stem cell precursors, undergo differentiation into adipocytes, also known as fat cells and the major component of adipose fat tissue (Loe, 2008). Muscle hypertrophy can be attributed by to both myogenic and adipogenic cell activation, although adipogenic progenitors are the only cells that assume adipogenic lineage. Other cells all have myogenic potential (Birbrair, et al, 2013).

Testosterone can stimulate pluripotent cells to differentiate and undergo myogenesis and can inhibit adipogenesis (Nieschlag, et al, 2012). Testosterone can inhibit adipogenic differentiation and therefore reduce a person's body fat percentage by influencing catecholamine signal transduction in fat cells, and by doing so is a key regulator of lipolysis, a mediator of the catabolism of cellular fat stores (Lass, et al, 2011). Two adipogenic inhibitory factors, PPARδ2 (peroxisomal proliferator-activated receptor) and CCAAT enhancer binding protein have been shown to inhibit the uptake of lipids in adipocytes, stimulate the catabolism of fat stores in lipolysis and inhibit the differentiation of adipocyte precursor cells, therefore preventing progenitor cells from assuming adipogenic lineage (Kadi, 2008).Therefore in the case of AAS users, when a large amount of testosterone is administered, the fact that testosterone has an inhibiting effect on adipogenesis and the formation of fat tissue and it stimulates muscle development and myogenesis, could potentially be a reason as to why AAS users experience a large amount of fat loss combined with huge gains in muscle mass whilst still consuming a calorie dense diet and in combination with weight training exercises.

Satellite cells are the major source of progenitors required in the regeneration and development of adult muscle tissue and are solely committed to the myogenic lineage in skeletal muscle (Birbrair, et al, 2013). Testosterone can stimulate the activity of satellite cells in myoblast culture systems (Kadi, 2008) by interacting with AR which both regulates and stimulates the differentiation of satellite cells, the main site of AR expression and other mesenchymal precursor cells in skeletal muscle tissue (Sinha, et al, 2004). Satellite cells, also known as myogenic precursor cells, remain quiescent in adult skeletal muscle in their satellite pool between the basal lamina (Brownell, et al, 1980) and the sarcolemma, plasma membrane of muscle fibres (Rainer, 2008).

Figure 21: the effect of satellite cells on skeletal muscle

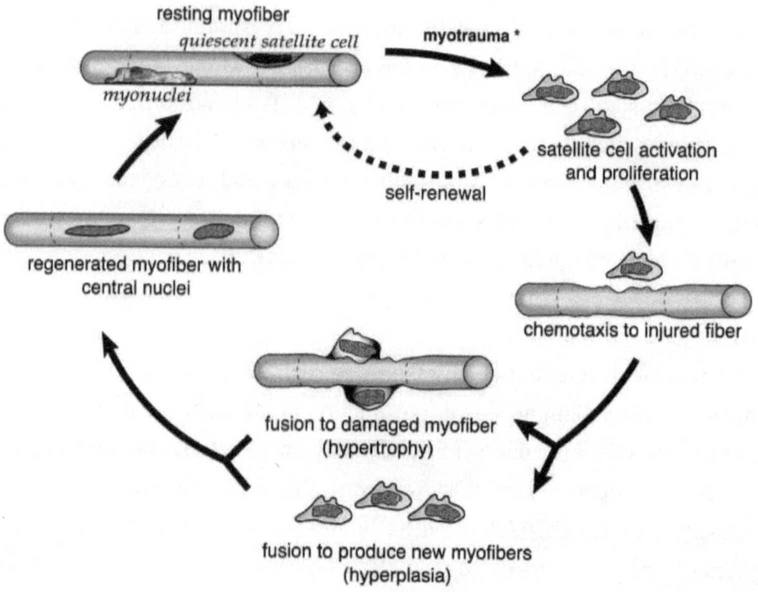

Figure 21: In response to a myotrauma brought about by weight training for example, satellite cells of the skeletal muscle become activated and begin proliferating. Some of the satellite cells will undergo self renewal and return to quiescence to replenish the satellite pool. The other satellite cells migrate to the muscle that is being put

under stain and undergoing catabolism. Depending on the severity of the workout and the amount of damage that the muscle has been subjected to, satellite cells may align and or fuse with myofibres in order to generate new myofibres. In this case, the nuclei of the fused satellite cell will migrate to a more peripheral location (Hawke, et al, 2001).

Satellite cells also have the ability to become activated and undergo proliferation in response to a stimulus such as weightlifting, to either lead to myonuclear accretion (gradual increase in myonuclei number), a large determinant of muscle growth potential (Mozdziak, et al, 1997) or lead to the formation of new myotubes, by fusing together with myofibres and donating their nucleus to the fibres. However they can also stop proliferating and return to quiescence in order to replenish the muscle stem cell satellite cell pool by generating new satellite cells. This may occur when satellite cells are forced to enter the cell cycle and some satellite daughter cells may escape differentiation after proliferation. However, it's not yet understood how the fate of the proliferating satellite cells are determined or regulated.

Satellite cells are however essential in the hypertrophy of skeletal muscle, as they represent the only way by which new myonuclei can be provided to muscle fibres (Verdijk, 2014; Kadi, 2008).

Testosterone can interact with AR which may directly affect the myonucleus of muscle cells. In a recent study conducted by Kvorning, et al (2014) 8 weeks of strength training dramatically increased the myonuclear number of type 2 muscle fibres (figure 22) by 12% in one of the study participants. In the same study, a second participant was treated with a GnRH analogue, goserelin, a testosterone suppressor which reduced the persons' resting testosterone levels to 10-20 times lower than that of the other study participant. In the case of the subject treated with goserelin, there was no change in myonuclear number after the 8 weeks of strength training.

Figure 22: Effects of testosterone on the myonuclear numbers in different muscle fibres

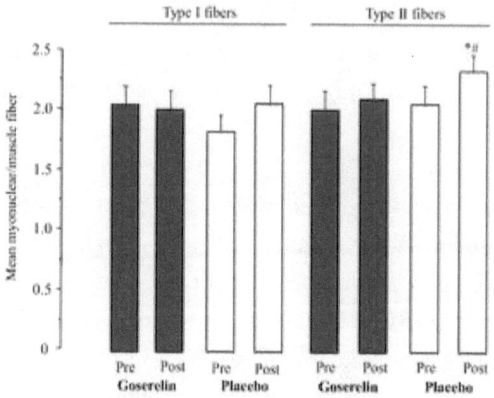

Figure 22: The effects of goserelin on type one and two muscle fibres pre and post 8 weeks of weight training (Kvorning, et al, 2014).

This study goes some way in showing that testosterone is a vital component of muscle hypertrophy and it's interactions with mesenchymal precursor cells in skeletal muscle tissue are essential in leading to an increase in myonuclear numbers upon stimulation by strength training and ultimately leading to an increase in lean muscle mass (Kvorning, et al, 2014).

Testosterone has the ability to enhance the transcriptional activity of the original myonuclei, which when it reaches its maximum level, can lead to increased protein synthesis and muscle hypertrophy (Kadi, 2008). The cellular mechanisms by which AAS interact with cellular machinery to cause protein synthesis, are demonstrated in figure 23:

Figure 23: the cellular mechanisms of AAS

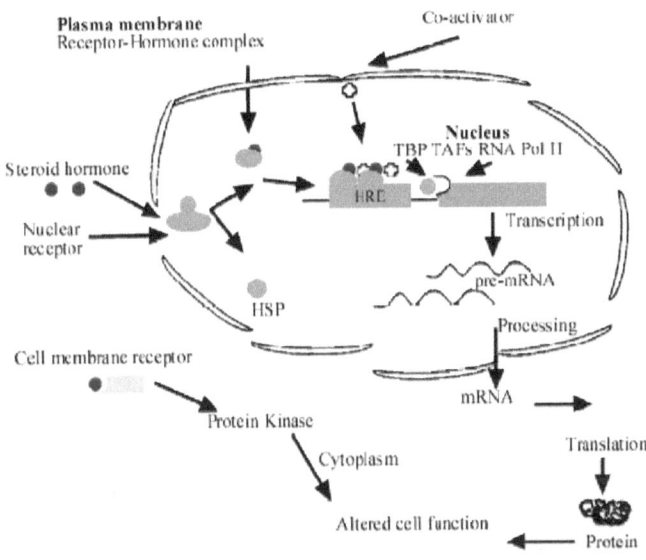

Figure 23: The cellular mechanisms of action of AAS, testosterone or the more potent DHT. AAS bind the AR, forming a receptor complex in the target cell. This causes heat shock protein (HSP) to detach. Translocation occurs and the complex binds HRE (hormone response element transcription regulators). Direct contact between the AR complex and the transcription factors enables transcription and translation to occur. This may also occur due to indirect action of coregulators. Transcription factors include the TATA Binding Protein (TBP), the associated TAFs (TBP associated factors) and RNA polymerases. The receptor complex translocates to binding sites on the chromatin which cause gene transcription and mRNA synthesis and processing. mRNA translation then takes place which results in protein synthesis. Steroids may act on GPCRs such as the GPRC6A receptor which has an important non-genomic regulatory function on androgens in different tissues, (Pi, et al, 2010; Gradisnik, et al, 2009). There may also be an AR that exists as a cell membrane receptor, although this has not yet been definitively proven. Non-genomic activity via interactions with cell membrane receptors may induce 2nd messenger signalling cascades which may cause an increase in intracellular calcium levels and activate protein kinases such as PKA or PKC, and the MAPK pathway which may lead to a range of different cellular events which may affect protein synthesis (Li, et al, 2009).

There are a number of modifications that give AAS their specific properties, including increasing their binding affinity to the AR. For example a steric substitution at the 7 α position can enhance affinity. Removing the 19-methyl group and 7α methylation of 17 hydroxy-4-androstenes can also enhance the affinity and the androgenic effects of the AAS (Fragkaki et al, 2009). Figure 24 depicts some additional sites of testosterone that may be the target for synthetic modification for AAS production.

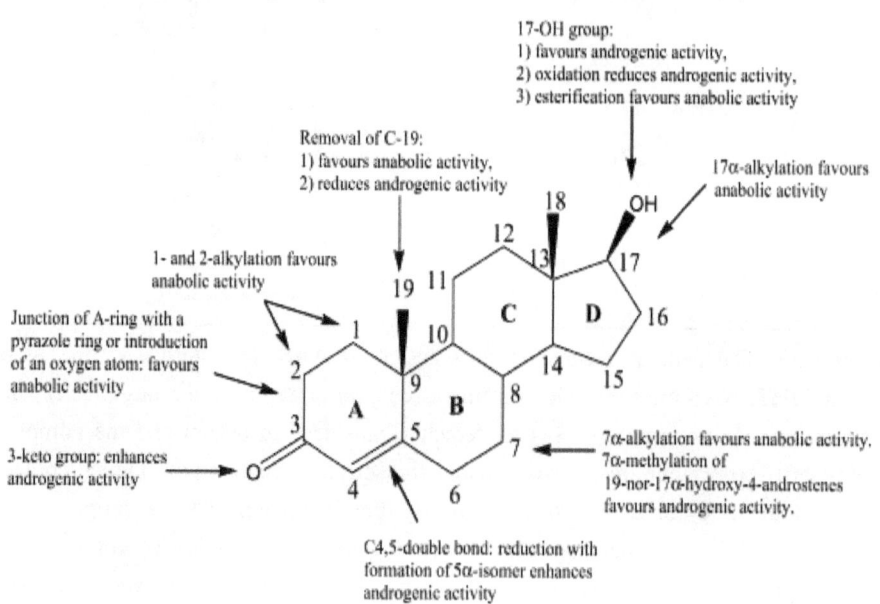

Figure 24: Targets for structural modification of testosterone to AAS

(Fragakaki et al, 2009).

It's possible that AAS and testosterone elicit their effects via GPCR mechanisms, although not much is currently known about this possible pathway. Figure 14 shows the relationship between testosterone, AAS and calcium. AAS increase intracellular calcium levels more than endogenous testosterone; this may involve a GPCR mechanism as opposed to the AR. This calcium increase is sensitive to pertussis toxin, a GPCR inhibitor, suggesting that the AAS and testosterone are binding a GPCR or a protein of similar structure and properties (Dubois et al, 2012). However this is yet to be definitively proven.

5 Side effects of AAS use

5.1 Cardiovascular effects

Much of the research that has been carried out into the abuse of AAS points to the fact that they can cause a number of cardiovascular problems (Riezzo, et al, 2011).

In a recent study conducted by Hassan, et al (2013), the effects of AAS on the process of angiogenesis, cardiac cell apoptosis and the histology of cardiac muscle was investigated. Using rats as a study group, it was found that AAS significantly increased both systolic blood pressure and heart rate, in comparison to those that were not subjected to AAS and the initial values of the group of rats that were investigated. The group of 40 rats were divided into groups of 10, with one group being the control group, one being subjected to AAS, another was exercise trained and the final group were both exercise trained and given AAS. The steroid used in this case was nandrolone decanoate (ND).

ND is one of the most widely used AAS amongst those looking for a drastic increase in both strength and muscle mass. It has been proven as an extremely effective agent in building body mass and fat free mass (FFM) in male bodybuilders who use a regular weight training program. This is demonstrated in this study (table 1) in which ND was injected intramuscularly at a dose of 200 mg a week for an 8 week period (van Lichtenbelt, et al, 2004).

Table 1: Effects of nandrolone on body composition

TABLE 1. Physical characteristics, and body composition data at baseline, after 8 wk of nandrolone decanoate (ND) or placebo treatment, and at 6 wk after completion of the intervention.

	Baseline			8 wk			14 wk			ANCOVA	
	Mean	SD	Range	Mean	SD	Range	Mean	SD	Range	Baseline–week 8 (P)	Baseline–week 14 (P)
Age (yr)											
ND	32.7		19.0–44.0								
Placebo	31.0		20.0–42.0								
BM (kg)											
ND	76.0	12.2	54.5–94.6	78.2*	11.9	57.6–96.3	77.6*	12.5	56.4–97.9	<0.05	NS
Placebo	84.1	8.9	69.6–96.6	84.6	9.1	69.2–97.9	84.2	8.8	70.4–97.2		
BMI (kg·m^{-2})											
ND	24.7	2.3	21.0–27.6	25.5*	2.1	22.2–28.1	25.2*	2.3	21.8–28.6	NS	NS
Placebo	26.8	2.0	23.8–29.8	26.9	2.1	23.7–30.2	26.8	2.2	24.1–30.0		
FFM (kg)											
ND	63.1	9.6	47.1–78.4	65.7*	9.7	50.0–80.7	65.0*	10.1	47.4–80.3	<0.05	NS
Placebo	68.2	6.3	57.2–74.5	69.0	6.6	57.6–74.7	69.2	6.1	59.9–75.5		
FM (kg)											
ND	12.9	3.9	7.4–18.8	12.5	3.6	6.7–16.9	12.6	3.7	7.4–17.5	NS	NS
Placebo	15.8	3.4	12.2–22.0	15.6	4.3	10.1–23.2	15.0	4.4	9.1–21.7		
%BF											
ND	16.8	3.6	12.0–23.2	15.8	3.5	8.9–21.1	16.1	3.4	10.0–21.0	NS	NS
Placebo	18.7	2.5	14.8–22.8	18.3	3.7	12.1–23.7	17.6	3.9	11.2–22.3		

BM, body mass; BMI, body mass index; FFM, fat-free mass; FM, fat mass; %BF, percentage body fat.
* Significant difference compared with baseline values (paired t-test).

Table 1: After 8 weeks, body mass increased in the ND group from 76.0kg to 78.2kg (an increase of 2.2kg) whereas in the placebo group body mass only increased from 84.1kg to 84.6kg (an increase of 0.5kg). This may be due to psychological effects; the placebo may have made the participants believe they were on ND which may have made them workout with more intensity during this period. T his may explain the slight increase in body mass also seen in the placebo group (van Lichtenbelt, et al. 2004).

In the study conducted by Hassan et al (2013), the rats were exercise trained by jumping into water and swimming in a tank for 30 minutes a day, 5 times a week for a period of 5 weeks. The regime consisted of 4 sets of 10 jumps. In order to simulate the weight training and exercise regimes carried out by those most likely to use AAS, in the second week, a load that was 50% of the rats bodyweight was strapped to its chest whist it was carrying out the same exercises. This load was gradually increased over the weeks, until in the final two weeks of the study, the rats were exercising with a load equivalent to 70% of their bodyweight. The groups receiving ND, received a dosage of 5mg/kg body weight, injected intramuscularly twice a week. Systolic blood pressure and electrocardiogram recordings (ECG) were taken at regular equal intervals throughout the study. Upon the conclusion of the study, blood samples were taken and tissue samples were prepared (Hassan, et al 2013).

Results indicated a rise in systolic blood pressure in the groups given AAS. The greatest rise was seen in the AAS group alone (an average rise of 40 mmHg in 6 weeks) (Hassan, et al 2013) (Table 2). There was also an increase in heart rate (HR) in the steroid group alone (Table 3); in the steroid plus exercise trained group, a significant elevated HR was not observed over the 6 weeks of the study, which could be attributed to the fact that exercise and physical activity does play a crucial role in reducing the resting HR by increasing the parasympathetic activity and decreasing sympathetic activity in the heart at rest (Carter, et al, 2003). This is evident from observing the reduction in resting HR of the exercise trained group over the 6 week period. Therefore exercise may possibly combat the effect that steroids have on HR by keeping it at a relatively stable rate. In the steroid group alone, a drastic increase in HR was seen, which may have occurred in order for the heart to be able to cope with the increase in body weight that was also determined in the study (Table 4).

Table 2: Systolic blood pressure (mmHg) in the 4 study groups

Groups (n=10)	1st week	2nd week	3rd week	4th week	5th week	6th week
Control	100±4.7	101±5.5	103±5.8	102±5.5	105±4.6	104±4.7
Steroid	107±4.2	112±6.4	120±8.6	129±7.1 **, #	142±8.2 ***, ##	147±9.3 ***, ##
Exercise-trained	104±5.3	110±7.2	112±7.5	116±6.2	115±5.5	112±9.2
Trained plus steroid	105±4.9	114±7.4	115±6.6	119±6.5	122±6.3 *, #	124±7.4 *, #

Data are presented as mean ± SEM. * $P<0.05$, ** $P<0.01$, *** $P<0.001$ versus control group. # $P<0.05$, ## $P<0.01$ versus first week.

Table 2: There were significant increases in systolic blood pressure in both the steroid and steroid and exercise trained groups which became evident from the 4th week onwards in comparison to the control group and the 1st week of the study. (Hassan et al, 2013).

Table 3: Heart rate (beats/min) in the 4 study groups

Groups (n=10)	1st week	2nd week	3rd week	4th week	5th week	6th week
Control	410±27	433±28	421±33	461±28	452±29	466±34
Steroid	422±29	505±35	631±36 ***, ##	677±38 ***, ###	793±43 ***, ###	851±50 ***, ###
Exercise-trained	405±25	326±19 **, #	311±18 **, ##	279±31 ***, ###	301±21 ***, ##	260±23 ***, ###
Trained plus steroid	401±24	386±26	391±22	485±34	474±39	492±42

Data are presented as mean ± SEM. ** $P<0.01$, *** $P<0.001$ versus control group. # $P<0.05$, ## $P<0.01$, ### $P<0.001$ versus first week.

Table 3: Heart rate was taken using an ECG, 24 hours after the last training session. The rats in the steroid group had an increased heart rate throughout the duration of the study that was significantly higher in the 6th week in comparison to the 1st week and the control groups at the same time period (Hassan et al, 2013).

Table 4: Body weight (BW), Heart weight (HW), HW/BW ratio and intraabdominal fat in each of the 4 study groups (Hassan, et al. 2013)

Groups	Body weight (g) Weak 1	Body weight (g) Weak 6	Heart weight (g)	HW/BW Ratio (mg/g)	Intraabdominal fat (g/ 100g BW)
Control (n=10)	219±4.2	395±8.2 §§§	1.13±0.05	2.86	5.18±0.63
Steroid (n=10)	223±5.1	351±6.6 ***, §§§	1.45±0.07 **	4.13 **	3.16±0.42 *
Exercise-trained (n=10)	227±4.5	335±7.7 ***, §§§	1.24±0.04 #	3.7 *	4.29±0.53
Trained plus steroid (n=10)	222±3.9	324±6.9 ***, #, §§§	1.30±0.04 *	4.01 *	3.65±0.23 *

Data are presented as mean ± SEM. * $P<0.05$, ** $P<0.01$, *** $P<0.001$ versus control group. # $P<0.05$ versus steroid group. §§§ $P<0.001$ versus weak 1.

Table 4: Each of the groups showed an increased BW over the 6 weeks and cardiac hypertrophy and hypertrophy of cardiomyocytes was significant in the exercise trained plus steroid groups (Hassan et al, 2013).

Exercise can bring about numerous benefits to cardiovascular health, although combined with AAS use, a number of disorders can become common place. AAS can elevate systolic blood pressure levels, perhaps due to salt (restricting dietary salt to 80mmol a day can reduce systolic blood pressure by 4.3mm Hg) and fluid retention, although the specific reasons are yet to be definitively determined (Bomze et al, 1991). However the fact remains that a rise in systolic blood pressure is a huge risk factor in the development of cardiovascular disease (Basile, 2002). Further studies on human subjects taking AAS have also shown a rise in systolic blood pressure (Riebe et al, 1992) (Table 5).

Table 5: mean blood pressure

	AS Users (N = 9)	Nonusers (N = 10)	Controls (N = 10)
SBP Rest (mm Hg)	130.1	124.9	120.1
DBP Rest (mm Hg)	78.8	80.5	80.4
Max TM SBP (mm Hg)	207.2	190.9	193.6
Max TM DBP (mm Hg)	87.2	78.9	84.1
Leg press SBP (mm Hg)	185.4	175.8	170.8
Leg press DBP (mm Hg)	104.5	97.5	98.2

Table 5: Results from a study conducted by Riebe et al (1992) into the blood pressure levels of human subjects self administering AAS and non AAS users. SBP=Systolic blood pressure, DBP=diastolic blood pressure, TM=treadmill. A rise in systolic blood pressure was measured in AAS users at rest and during various exercises (Riebe et al, 1992).

Systolic hypertension is due to increased arterial stiffness and is characteristic of an elevation of systolic blood pressure to levels of >140 mm Hg, which puts those in the steroid group in table 2 at risk (Franklin, 2004). Systolic hypertension correlates with an increased heart rate (table 3) and in worst case scenarios, can lead to heart failure and death.

Long term AAS use has been known to raise levels of low density lipoprotein cholesterol and lower high density lipoprotein cholesterol that carry cholesterol in the blood, factors associated with coronary heart disease and also a factor in the development of cardiovascular disease (Bomze et al, 1991). However this change in lipoprotein levels is dependent on the type of steroid used and the length of a steroid cycle. Stanozolol for example, taken orally at a dose of 6mg a day for a 6 week cycle, reduces HDL cholesterol levels by 33%, whereas testosterone ethanthate taken by the participants at a weekly dose of 200mg for the same 6 week cycle reduced HDL by 9%. These changes in lipoprotein levels are a great risk factor in the development of atherosclerosis. Other risk factors associated with ASS use include blood clotting, myocardial hypertrophy and arrhythmias. Many of these effects such as hypertension are reversible and can improve when AAS use is stopped. However there are many, such as atherosclerosis and cardiomyopathy that seem to be irreversible (van Amsterdam et al, 2010).

There have been a number of reported incidences of high profile athletes or others that have died or have had health complications that have been attributed to cardiovascular disease and AAS use. A study conducted by Fineschi, et al (2007), investigated whether the sudden cardiac death of two male bodybuilders was due to steroid abuse. These men before their death were stacking using several different types of oral and injectable AAS, including: nandrolone, stanozolol, methenolone and many others. Upon their deaths, Fineschi et al (2007) were able to carry out an autopsy, histology of the organs (figure 25) and toxicological analysis. Through their findings they claimed that the deaths of the two bodybuilders were due to AAS abuse (Fineschi, et al, 2007).

Figure 25: Histological findings from the study conducted by Fineschi et al (2007) upon the death of 2 AAS users

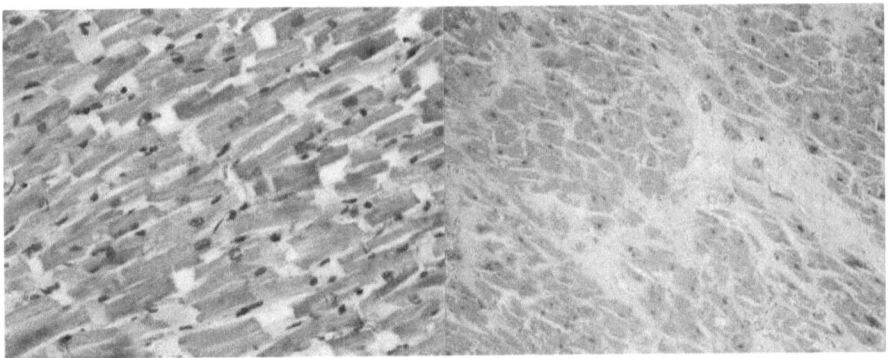

Figure 25: EM microscopy shows widening of the intercallated disks and granular disruption (left diagram) in the first AAS user. This may reduce action potential upstroke velocity due to low sodium current density. This would lead to reduced conduction rates and would increase the risk of arrhythmias (Rizzo et al, 2012). The second AAS user (right diagram) showed signs of scar fibrosis (Fineschi et al, 2007) which can cause cardiac diastolic dysfunction and a range of other heart conditions, ultimately leading to cardiac death.

However it still cannot be definitively said whether these problems are attributable to AAS alone. Since very little is known about these patients prior to their deaths, it cannot be said whether the individuals were taking other drugs, abusing other substances, taking AAS and stacking in a continuous cycle or on and off, using high dosages and for what period of time. There are too many variables in an individual's lifestyle for it to be claimed that death is due to AAS alone or a certain type of AAS.

There is also no epidemiological evidence that cardiovascular disease is due to the abuse of AAS and the majority of the evidence that has been gained from studies and research is merely circumstantial. In addition to this, many AAS users have at one point or other experimented with other drugs such as human growth hormone or EPO (erythropoietin) and so the risks of cardiovascular complications associated with AAS use cannot be attributed to AAS alone (van Amsterdam et al, 2010).

5.2 Immune system

It has been suggested that supraphysiologic doses of AAS may cause a greater amount of natural killer activity such as T cell proliferation in those AAS with an altered steroid nucleus and cause lower immunoglobulin levels such as IgA, which can cause autoimmune tissue damage and an increased risk to certain infections (Bomze et al, 1991). This is supported by Gradisnik et al (2009), who reported that AAS can reduce the number of immune cells and have adverse effects on their function. AAS may also have an effect on lymphocyte differentiation, antibody and cytokine production and so therefore have an adverse effect on immune function and activity. T lymphocytes possess AR at their immature stage development which provides evidence that T cells and AAS may be associated with each other. However this is yet to be definitively proven, and further studies conducted by Calabrese et al

(1989), as shown in Table 6, have found no significant differences in lymphocyte subpopulation counts between AAS and non AAS users.

Table 6: Lymphocyte subpopulations in AAS and non AAS users

Variable	Group	N	Mean	SD	Median	Min	Max
Total lymphocytes							
	Steroid users	13	1,767.4	492.9	1,672	1,020	2,800
	Non-users	10	1,537.8	513.2	1,470	798	2,508
	Controls	20	1,844.8	437.4	1,710	1,160	2,924
T-cells							
	Steroid users	13	1,222.3	389.0	1,121	806	2,178
	Non-users	8	1,091.6	314.4	1,026	610	1,573
	Controls	20	1,364.0	316.4	1,296	867	1,958
T-helper/inducer							
	Steroid users	13	865.0	342.8	777	521	1,792
	Non-users	8	717.5	226.4	657	384	1,098
	Controls	20	914.9	239.4	916	529	1,374
T-cytotoxic/suppressor							
	Steroid users	13	500.2	122.0	532	247	641
	Non-users	8	523.9	231.4	435	278	988
	Controls	20	523.9	179.0	499	236	994
HLA-DR positive cells							
	Steroid users	11	73.6	29.8	74	44	128
	Non-users	5	74.6	31.4	70	43	108
	Controls	20	88.5	40.0	82	42	183

All values are in cells per mm^3.

Table 6 - Shows the differences in the total lymphocyte count, T-cells, T-helper cells, T-suppressor cells and HLA-DR positive cells in AAS users, non users and a control group consisting of non bodybuilders. There were no significant differences between any of the groups (Calabrese et al, 1989).

It is difficult so say conclusively whether AAS are solely responsible for these issues. Since the majority of users use a different mix of AAS including orals and injectables, are on different cycles and may pyramid with varying dosages of each AAS, the effects of long terms use of AAS on the immune system remains subject to speculation (Gradisnik et al, 2009). This is the case for the study conducted by Calabrese et al (1989), the data of which is represented in Table 6. The steroid group in this study self administered a range of drugs including stanozolol, oxandrolone, nandrolone decanoate and phenpropionate and many others in varying dosages and so the results cannot be attributed to any single AAS. There are also a lot of conflicting views as to whether AAS do in fact damage the immune system or enhance antibody production and natural killer cell activity in response to immune damage (Hughes et al, 1998). Figure 26, although using a small sample size, demonstrates a significant difference in the natural killer activity between AAS and non AAS users.

Figure 26: Natural killer activity in AAS and non-steroid users

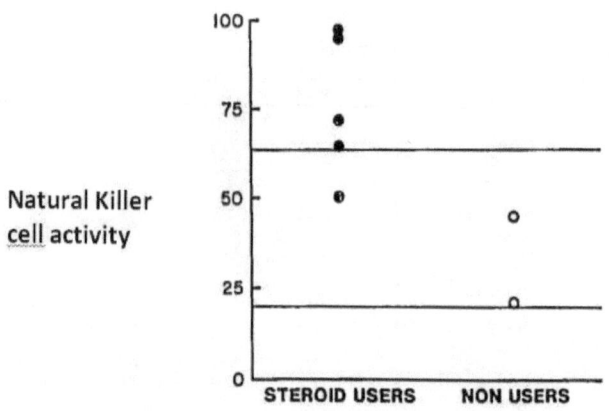

Figure 26: Natural killer cell activity in 5 AAS and 2 non-steroid users. 4 out of the 5 AAS users had figures above the predetermined normal range of 20-66% (Calabrese et al, 1989).

AAS are also used to treat a variety of immune deficiency syndromes such as HIV, although when used for this purpose their main role is in the prevention of HIV–related wasting myopathy and not to boost the immune system, although HIV patients on steroids have reported feeling an increased sense of overall general well being (Berger et al, 1993). There are a number of conflicting views as to whether AAS can adversely effect or benefit immune responses, but the fact remains that more research need to be carried out where, unlike the data in figure 26, dosages and AAS cycles are monitored and stacking is kept to a minimum, so that the specific effects of a certain AAS on immune responses can be determined.

5.3 Liver

Effects on the liver are common amongst AAS users. Side effects include an elevated concentration of liver enzymes, such as SGOT (Serum glutamic oxaloacetic transaminase), SGPT (serum glutamic-pyruvic transaminase) and LDH (lactate dehydrogenase). In rare cases, hepatitis and jaundice has also occurred. The specific mechanisms as to why AAS cause damage to the liver is yet to be determined, but it has been hypothesised that AAS have a toxic effect on the bile secretory mechanisms (Elsharkawy et al, 2012). This may result in the failure to secret and hence breakdown certain substances by the liver, which may result in liver injury or failure. However there is little evidence currently available to support this hypothesis. AAS effects on the liver tend to be more common in those that incorporate a large amount of oral steroids in their cycles. Oral AAS are mainly 17-α alkylated derivatives of testosterone, that have undergone alkylation at the 17α position with an ethyl or methyl group. This structural modification makes oral AAS intolerable to liver degradation and increases the drugs half life meaning the AAS have a longer lasting effect in the body and exposing hepatocytes and cholangiocytes to the drug for a longer time period (Elsharkawy et al, 2012). This intolerance to liver degradation is mainly due to the methyl group which inhibits the active site of the enzymes, thus making oral AAS more toxic to the liver. Injectables are 17-β ester derivatives that have undergone esterification at the 17-β hydroxyl group. Due to this modification, injectable AAS have an increased lipid solubility and slow release and rate of absorption blood circulation (Gradisnik et al, 2009).

Oral AAS such as stanozolol are thought to be responsible for liver damage, mainly due to the fact that injectables (Osorio et al, 2008) and non-17-alkylated anabolic steroids have failed to produce abnormal liver function tests (Marquardt et al, 1964). However, more research has to be carried out as to whether AAS are in fact the cause of these elevated enzyme levels because intense weight training can also elevate enzyme levels, so more specific forms of blood analysis has to be carried out to differentiate the sources of these changes (Bomze et al, 1991).

There have been reported incidences of both benign and malignant tumours in the liver in individuals using AAS over several decades, although these may regress if AAS use is stopped permanently (Bomze et al, 1991).

Oral AAS may also induce peliosis hepatis, in which blood filled sacs in the liver can rupture, leading to liver failure and possibly death, (Bomze et al, 1991) although these are rare isolated cases and whether AAS alone are the cause of these illnesses are yet to be determined. However in one such case, in which an individual had been reported to have had jaundice that the researchers hypothesised was brought about due to AAS use (the individual later admitted to taking 5 mg of methandrostenolone a day, however the duration of the cycle was unknown), a liver biopsy was taken (figure 27), which showed cholestasis and the extent of liver injury that the researchers stated was due to AAS use.

Figure 27: Liver biopsy of a patient using AAS

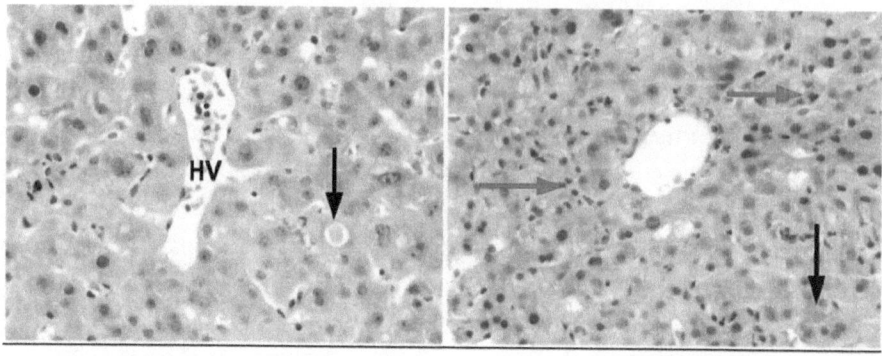

Figure 27: The figure on the left shows cholestasis and the arrow shows canalicular bilirubinostasis (retention of bilirubin in the canalicular) which can progress to cause jaundice, and shows inflammation in the hepatic vein (HV). The diagram on the right shows symptoms of cholestatic liver disease with hepatocellular bilirubinostasis (shown by the vertical arrow) and cholestatic hepatitis (red arrows) (Elsharkawy et al, 2012).

This demonstrates that AAS can cause liver injury, the extent of which is still unknown. Since the patient that had the liver biopsy shown in figure 27 categorically denied the fact that he was on AAS (after a substantial amount of further questioning he admitted to being a AAS user), these liver injuries may be a consequence of stacking and a prolonged period of usage, and so further research needs to be carried out in order to determine the effects AAS have on liver function.

5.4 Infertility

When AAS are administered to patients in a medical setting, they are given doses that would be equivalent to the body's natural testosterone production range. However, when users take AAS for recreational use, many take up to 100 times of this recommended amount. Consequently, the body's natural testosterone production stops. In addition to a decrease in endogenous testosterone levels, FSH and FH levels may also decrease when on AAS (Bomze et al, 1991). This is because if the body is already receiving AAS, which in most cases are being directly injected into the muscle to act directly on target tissues, the body can bypass the mechanisms leading to testosterone production, e.g. GnRH no longer needs to stimulate LH production and therefore LH no longer stimulates testosterone production in the testis. The large amounts of exogenous testosterone causes a negative feedback loop that suppresses the production of GnRH from the hypothalamus.

Figure 28: Possible negative feedback inhibition loop, leading to a decreased amount of endogenous testosterone production in AAS users

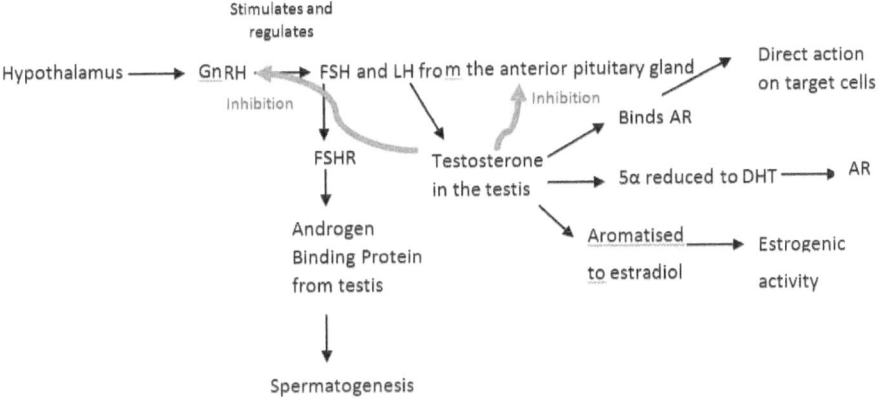

Figure 28: GnRH is secreted from the hypothalamus, which stimulates and regulates the production of FSH and LH from the anterior pituitary gland. FSH binds to its receptor which stimulates the production of androgen binding protein in the testis. This stimulates spermatogenesis. LH can stimulate testosterone production in the testis which can either bind to an androgen receptor and directly act on target cells, can be 5α reduced to DHT which can then bind AR, or it can be aromatised to estradiol which could lead to estrogenic activity. If testosterone is administered in the form of AAS, it may cause a negative feedback inhibition loop; FSH and LH production and secretion may be inhibited from the anterior pituitary gland or direct inhibition may occur at the hypothalamus to suppress GnRH secretion.

A result of these inhibitory effects may be testicular atrophy, a condition which is consistent with the suppression of the hypothalamic–pituitary–gonadal axis by exogenous anabolic steroids, which could also cause secondary hypogonadism. Inhibition of the secretion of GnRH is the primary cause of testicular atrophy. AAS may also decrease sperm production due to its inhibiting effects on FSH which causes spermatogenesis. The hypothalamic– pituitary axis may return to normal or may only show signs of partial recovery, which may take months or years and is dependent on the dosages taken and the amount of time that AAS were used (Boregowda et al, 2011). Due to these hormonal changes and testicular atrophy, AAS can cause a persistent impairment on testicular and male reproductive functions and therefore cause male infertility (de Souza et al, 2011).

5.5 Additional side effects

There are numerous other side effects brought about by AAS use. People respond differently to AAS so many side effects will vary from person to person.

Side effects are also largely dose dependent, with those taking larger doses of AAS for a longer amount of time more likely to experience one or several different side effects.

Certain users may also stack with several different AAS, some of which may enhance a certain side effect.

Many of the undesirable side effects can be attributed to the androgenic masculinising effects of the drugs, such as hair loss and deepening of the voice. Table 7 shows some of the possible side effects associated with AAS usage.

Table 7: Possible adverse effects associated with AAS use

Target	Adverse effect	Description
Bone	Premature closure of the epiphysis in children	Stunting of linear growth
Breast	Gynaecomastia and enlarged nipples in men	Gynaecomastia can be pronounced and painful; corrective surgery may be necessary. As some anabolics are known to be resistant to aromatisation, other mechanisms need to be considered, such altered hepatic function causing an imbalance between androgens and estrogens

Target	Adverse effect	Description
Cardiovascular	Increases risk of thrombotic events such as myocardial infarction or stroke (raised LDL, lowered HDL and apolipoprotein-1, raised haematocrit (due to polycythaemia) and lowered plasma fibrinogen Cardiac damage (left ventricular hypertrophy, fibrosis and heart failure). Sudden cardiac death	The link between long-term anabolic steroid use and cardiovascular events remains to be clearly established but the evidence gathered is fairly compelling. This effect is probably underreported Heart disease may be potentiated by concomitant use of growth hormone or insulin (also misused for anabolic purposes)
CNS	Hypomania (less severe form of mania) Heightened irritability Increased aggression and hostility Destructive impulses Self-destructive impulses Withdrawal symptoms can include severe depression	Psychological effects are unpredictable. Anabolic steroids are implicated in cases of violent behaviour ('roid rage') including, manslaughter and murder Polypharmacy can increase the risk of violent criminality
Hair	Acceleration of baldness in men; male-pattern baldness in women.	Hirsutism is at very best only partially reversible on cessation of administration.
Liver	Impaired function Hepatic cholestasis (bile canal obstruction) causing jaundice Peliosis hepatitis (blood-filled sacs in the liver) Liver tumours (increased risk)	Hepatoxicity is associated with 17α-alkylated steroids (orally active) For the liver function test, it is important to include GGT and CK, as ALT and AST can be raised naturally due to exercise
Reproductive	Increased libido in men and women, which may be difficult to control Suppression of gonadal steroidogenesis Amennorhoea Testicular atrophy Disproportionate growth of the inner prostate	Recovery of fertility can take from months up to approximately 1 year after cessation of administration, depending on the extent of abuse
Skin	Cystic acne	This can be very severe on the chest, back and face

Target	Adverse effect	Description
Vocal chords	Deepening of the voice	Irreversible deepening of the voice can result in considerable distress
Other	Serious infection associated with injected drugs	Needle-exchange programmes are helping to address this problem
	Possible toxicity from unlicensed products	Extent is unknown

Table 7: Possible adverse effects associated with the use of AAS; ALT (alanine aminotransferase), AST (aspartate aminotransferase), CK (creatine kinase), GGT (γ-glutamyl transpeptidase), HDL (high-density lipoprotein), LDL (low-density lipoprotein) (adapted from Kicman, 2008).

Many of these effects can be difficult to recognise without a thorough medical examination, and doctor-patient confidentiality prevents these results from coming to light. Many of the androgenic side effects are noticeable, such as hair loss and deepening of the voice, but there are some androgenic and anabolic effects that are insidious such as the effects AAS may have on liver function, where the user may not realise that they have an illness until they begin to experience a certain level of discomfort. But the health risks and side effects associated with AAS use are dependent on a number of factors: the gender of the user, the dosages, the length of an AAS cycle, the different stacks being incorporated into a cycle, the amount of oral steroids being taken and 'the susceptibility of the individuals themselves to androgen exposure, which is dependent on genetics, age, and lifestyle' (Kicman, 2008).

6 Conclusion and future research

Many who use AAS do so in varying dosages and frequency, a term known as steroid cycles. These cycles vary between individuals, with those looking to increase muscle mass and those wanting to reduce body fat using a combination of different steroids, on and off, from anywhere from a few weeks to months at a time. It is therefore problematic when it comes to proving whether any detrimental effects are due to prolonged steroid cycles for years at a time or down to the combination of AAS that an individual uses in a cycle at any given time.

The effects of individual AAS have been determined, such as the effects of ND on BM and FFM as determined by van Lichtenbelt, et al (2004), but because users of AAS tend to use a combination of AAS, it is highly unlikely that any side effects will ever be attributed to the use of a single AAS.

AAS can be used in a safe, therapeutic way, however the doses that tend to be used by the vast majority of AAS users looking to increase muscle and strength, greatly exceed the maximum therapeutically recommended dose for AAS (Hartgens et al, 2004).

Although there have been studies of note that have been able to prove that the use of AAS do present an individual with certain side effect, the studies that have been carried out have only taken place on human subjects over a period of about 20 months, which is not a significant time frame in order for possible side effects to become apparent. There has only ever been rare isolated incidences of severe life-threatening side effects in AAS users and even these cases cannot be attributed to AAS alone and are down to severe abuse of the drugs rather than sustained moderated usage, and such side effects and incidences may not become apparent until 20 years or more of chronic and widespread AAS abuse (Lamb, 1984). Many people argue that this is where the problems stem from, because if a teenager wants to look bigger and more muscular in a month or so, the side effects that may or may not occur 20 years later are rarely a consideration. Studies into long term usage are unrealistic to conduct due to the number of ethical implications. It would not be ethically plausible to give human subjects AAS for a prolonged time period without being certain of the resulting medical implications of doing so whilst at the same time controlling all the variables in order to be able to measure and study the variables of choice. Of the individuals involved in studies relating to AAS, many have different health backgrounds, genetic makeup, exercise regimes and histories relating to AAS use. Even if the dosages of AAS given to subjects during a study are controlled, it is near on impossible to definitively claim that there is a cause–effect relationship between AAS and the side effects that may be associated with their use. The majority of studies carried out on the effects of AAS have been carried out using animals, mainly rats as study subjects. Although there have been notable effects determined from these studies, such as reduced life expectancy and a host of other medical problems, data derived from these animal studies are not necessarily transferable and comparable to the effects that AAS may have on humans (Sterngass, 2010).

As more research is carried out into the side effects and potential dangers of AAS use and various governments begin imposing greater restrictions on their possession and usage, those looking for a substance to aid muscle hypertrophy and strength gains may begin to look to using other substances such as plant brassinosteroids or human growth hormone (Esposito et al, 2011).

There is no doubt that AAS build muscle and strength, which is why there is a demand for AAS on the black market. Underground laboratories are constantly finding ways to produce undetectable AAS, and so detection methods are constantly being evaluated for their effectiveness. However for those that use AAS in a recreational sense, getting caught is rarely a serious consideration. Imposing tougher sanctions on AAS importation and use may go some way in restricting their use, but a greater amount of research is still needed in order to educate AAS users and potential users as to the possible side effects that are associated with AAS.

In conclusion, it has been well documented through numerous studies that testosterone causes skeletal muscle hypertrophy. AAS enhances hypertrophic properties and aid in fat loss resulting in large gains in muscle mass making these drugs extremely popular and the subject of ongoing research.

References

Adamski. J, Jakob. F. J; A guide to 17β-hydroxysteroid dehydrogenases, Molecular and Cellular Endocrinology, 2001, Vol. 171, Issues:1-2, pg. 1-4.

Aggarwal T, Polanco MJ, Scaramuzzino C, Rocchi A, Milioto C, Emionite L, et al. Androgens affect muscle, motor neuron, and survival in a mouse model of SOD1-related amyotrophic lateral sclerosis. Neurobiol Aging. 2014;35(8):1929-38.

Allen. D.G, Lamb. G.D, Westerblad. H; Skeletal Muscle Fatigue: Cellular Mechanisms, American Physiological Society: Physiological Reviews, 2008, Vol. 88, No.1, pg. 287-332.

Anderson K, Rosner W, Khan M, New M, Pang, S, Wissel P, Kappas A; Diet-hormone interactions: Protein/carbohydrate ratio alters reciprocally the plasma levels of testosterone and cortisol and their respective binding globulins in man, Life sciences, 1987, Vol.40, Issue 18, pg. 1761-1768

Awrejcewicz. J, Lind. Y.B, Kazakova. D.S; Modeling of sarcomere contraction kinetics, Differential Equations and Dynamical Systems, 2013, Vol.21, Issue:1, pg.15-9.

Bagatell. C. J, Bremner. W. J; Androgens in Men- Uses and Abuses, The New England Journal of Medicine, 1996, Vol.334, No.11, pg.707.

Basile J.N; Systolic blood pressure, BMJ, 2002, Vol.325, Issue:7370, pg.917-8.

Becker. KL; Principles and Practice of Endocrinology and Metabolism, 3rd edition, 2001, USA, Lippincott Williams and Wilkins, pg.957

Benardot. D; Advanced Sport Nutrition: 2nd Edition, 2012, USA, Versa Press.

Berger. J.R, Pall. L, Winfield. D; Effect of anabolic steroids on HIV-related wasting myopathy, Southern Medical Journal, 1993, Vol.86, Issue:8, pg.865-6.

Bhasin S, Woodhouse L, Storer T. Proof of the effect of testosterone on skeletal muscle. J Endocrinol. 2001;170(1):27-38.

Bhasin. S, Calof. O.M, Storer. T.W, Lee. M.L, Mazer. N.A, Jasuja. R, Montori. V.M, Gao. W, Dalton. J.T; Reproductive endocrinology (including placental hormones): Drug Insight: testosterone and selective androgen receptor modulators as anabolic therapies for chronic illness and aging, Nature Reviews Endocrinology, 2006, Vol. 2, pg. 146-159.

Bhasin S, Storer T, Berman N, Callegari C, Clevenger B, Phillips J, Bunnell T, Tricker R, Shirazi A, Casaburi R; The Effects of Supraphysiologic Doses of Testosterone on Muscle Size and Strength in Normal Men, The New England Journal of Medicine, 1996, Vol. 335, pg. 1-7.

Bhasin. S, Taylor. W.E, Singh. R, Artaza. J, Hikim. I.S, Jasuja. R, Choi. H, Cadavid. N.F.G; The Mechanisms of Androgen Effects on Body Composition: Mesenchymal Pluripotent Cell as the Target of Androgen Action, Journal of Gerontology: Medical Sciences, 2003, Vol.58, Issue: 12.

Birbrair. A, Zhang. T, Wang. Z-M, Messi. M.L, Enikolopov. G.N, Mintz.A, Delbono. O; Role of Pericytes in Skeletal Muscle Regeneration and Fat Accumulation, Stem Cells and Development, 2013, Vol.22, Issue: 16, pg. 2298- 2314.

Bomze. J.P, Cox. M.H, Miles. D.S; Anabolic Steroids: A historical and clinical perspective, National Strength & Conditioning Association Journal, 1991, Vol.13, Issue:42.

Boregowda. K, Joels. L, Stephens. J.W, Price. D.E; Persistent primary hypogonadism associated with anabolic steroid abuse, Fertility and Sterility, 2011, Vol.96, Issue:1.

Boulpaep. E. L, Boron. W.F; Medical physiology: a cellular and molecular approach, 2005, St. Louis, Mo: Elsevier Saunders. pg. 1125.

Brownell. A.G, Slavkin. H.C; Role of basal lamina in tissue interactions, Renal Physiology, 1980, Vol.3, Issue: 1-6, pg. 193-204.

Calabrese. L.H, Kleiner. S.M, Barna. B.P, Skibinski. C.I, Kirkendall. D.T, Lahita. R.G;The effects of anabolic steroids and strength training on the human immune response, Medicine and Science in Sports Exercercise, 1989, Vol.21, Issue: 4, pg.386-92.

Carter JB, Banister EW, Blaber AP; Effect of endurance exercise on autonomic control of the heart, Sports Med, 2003; Vol.33, Issue:1, pg.33-46.

Chen. H.H, Maeda. T, Mullett. S. J, Stewart. A.F; Transcription cofactor Vgl-2 is required for skeletal muscle differentiation, Genesis – The Journal of Genetics and Development, 2004, Vol.39, Issue:4, pg. 273-279.

Choong. K, Lakshman. K.M, Bhasin. S; The physiological and pharmacological basis for the ergogenic effects of androgens in elite sports, Asian Journal of Andrology, 2008, Vol.10, Issue: 3, pg. 351-363.

Chu. D.A, Myer. G; Plyometrics, 2013, USA, Human Kinetics.

Connective tissue of the skeletal muscle; image source:
http://www.tarleton.edu/Departments/anatomy/musclepix2.html

de Souza. G. L, Hallak. J; Anabolic steroids and male infertility: a comprehensive review, BJU International, 2011, Vol.108, Issue: 11, pg. 1860–1865.

Delavier. F, Gundill. M; How a muscle gains strengh, The Strength Training Anatomy Workout, 2011, France, Editions Vigot.

Dodet. B, Vicari. M; Pluripotent Stem Cells: Therapeutic Perspectives and Ethical Issues, 2001, Paris, John Libby.

Dubois V, Laurent M, Boonen S, Vanderschueren D, Claessens F. Androgens and skeletal muscle: cellular and molecular action mechanisms underlying the anabolic actions. Cellular and Molecular Life Sciences. 2012;69(10):1651-67.

Duchateau. J, Semmler. J.G, Enoka. R.M; Training adaptations in the behaviour of human motor units, Journal of Applied Physiology, 2006, Vol.101, Issue: 6, pg. 1766-1775.

Egan. B, Zierath. J.R; Exercise Metabolism and the Molecular Regulation of Skeletal Muscle Adaptation; Cell Metabolism, 2013, Vol.17, Issue: 2, pg. 162- 184.

Elsharkawy. A.M, McPherson. S, Masson. S, Burt. A.D, Dawson. R.T, Hudson. M; Cholestasis secondary to anabolic steroid use in young men, BMJ, 2012;344:e468.

Encyclopedia of Molecular Pharmacology, 2008.

Esposito. D, Komarnytsky. S, Shapses. S, Raskin. I; Anabolic effect of plant brassinosteroid, The Journal of The Federation of American Societies for Experimental Biology, 2011, Vol.25, Issue: 10, pg.3708-19.

Estrada M, Espinosa A, Müller M, Jaimovich E. Testosterone stimulates intracellular calcium release and mitogen-activated protein kinases via a G protein-coupled receptor in skeletal muscle cells. Endocrinology. 2003;144(8):3586-97.

Farina. D, Negro. F, Dideriksen. J. L; The effective neural drive to muscles is the common synaptic input to motor neurons, The Journal of Physiology, 2014, Vol. 592, Issue: 16, pg. 3427-3441.

Feher. J; Skeletal muscle mechanics. In: Elsevier Inc; 2012. p. 239-48.

Fineschi. V, Riezzo. I, Centini. F, Silingardi. E, Licata. M, Beduschi. G; Sudden cardiac death during anabolic steroid abuse: Morphologic and toxicologic findings in two fatal cases of bodybuilders, Int J Legal Med, 2007, Vol.121, Issue:1, pg.48-53.

Fragkaki. A.G, Angelis.Y. S, Koupparis. M, Tsantili-Kakoulidou. A, Kokotos. G, Georgakopoulos. C; Structural characteristics of anabolic androgenic steroids contributing to binding to the androgen receptor and to their anabolic and androgenic activities: Applied modifications in the steroidal structure, Steroids, 2009, Vol.74, Issue: 2, pg. 172-197.

Franklin. S.S; Systolic blood pressure, American Journal of Hypertension, 2004, Vol.17, Issue: 12.

Fry. A.C, Webber. J.M, Weiss. L.W, Harber. M.P, Vaczi. M, Pattison. N.A; Muscle Fibre Characteristics of Competitive Power Lifters, Journal of Strength and Conditioning Research, 2003, Vol.17, Issue: 2, pg.402-410.

Gard. PR, 2001; Human Pharmacology, London, Taylor and Francis.

Gondin. J, Cozzone. P.J, Bendahan. D; Is high frequency neuromuscular electrical stimulation a suitable tool for muscle performance improvement in both healthy humans and athletes?, European Journal of Applied Physiology, 2011, Vol. 111, Issue: 10, pg. 2473-2487.

Gradisnik. S.M, Green. R, Brenu. E.W, Weatherby. R.P; Anabolic androgenic steroids effects on the immune system: a review, Central European Journal of Biology, 2009, Vol.4, Issue:1, pg. 19-33.

Gunther. S, Kruse. K; Spontaneous sarcomere dynamics, Chaos AIP: an interdisciplinary Journal of Nonlinear Science, 2012.

Hackney, A.C. Testosterone and reproductive dysfunction in endurance-trained men. In: Encyclopedia of Sports Medicine and Science, T.D. Fahey (Editor). Internet Society for Sport Science: http://sportsci.org. 1998.

Hall. J. E, Schoenfield. D. A, Martin. K. A, Crowley. W.F Jr.; Hypothalamic gonadotropin-releasing hormone secretion and follicle-stimulating hormone dynamics during the luteal-follicular transition, The Journal of Clinical Endocrinology and Metabolism, 1992, Vo.74, Issue:3, pg..600-607.

Hall. R.C.W, Chapman. M.J; Psychiatric complications of anabolic steroid abuse, Psychosomatics. 2005, Vol.46, Issue: 4, pg.285-90.

Hartgens F, Rietjens G, Keizer HA, Kuipers H, Wolffenbuttel BHR; Effects of AAS on apolipoprotiens and lipoprotein (a), Br J Sports Med, 2004, 38:253-259: http://bjsm.bmj.com/content/38/3/253.long.

Hassan AF, Kamal MM. Effect of exercise training and anabolic androgenic steroids on hemodynamics, glycogen content, angiogenesis and apoptosis of cardiac muscle in adult male rats. International journal of health sciences. 2013;7(1):47-60.

Hawke. T.J, Garry. D.J; Myogenic satellite cells: Physiology to molecular biology, Journal of Applied Physiology, 2001, Vol.91, Issue:2, pg.534-51.

Horne. Z, Hesketh. J; Increased association of ribosomes with myofibrils during the skeletal-muscle hypertrophy induced either by the beta-adrenoceptor agonist clenbuterol or by tenotomy, The Biochemical Journal,. 1990, Vol. 272, Issue: 3, pg.831-3.

Hughes. T.K, Fulep. E, Juelich. T, Smith. E.M, Stanton. G.J; Modulation of immune responses by anabolic androgenic steroids, International Journal of Immunopharmacology, 1995, Vol.17, Issue:11, pg. 857-863.

Hughes. T.K, Rady. P.L, Smith. E.M; Potential for the effects of anabolic steroid abuse in the immune and neuroendocrine axis, Journal of Neuroimmunology, 1998, Vol.83, Issue: 1-2, pg. 163-167.

Ip. E.J, Barnett. M.J, Tenerowicz. M.J, Kim. J.A, Wei. H, Perry. P.J; Women and anabolic steroids: An analysis of a dozen users, Clinical Journal of Sport Medicine, 2010, Vol.20, Issue:6, pg.475-81.

Kadi, F; Cellular and molecular mechanisms responsible for the action of testosterone on human skeletal muscle. A basis for illegal performance enhancement. British Journal of Pharmacology, 2008, 154: 522–528.

Kanayama. G, Brower. KJ, Wood. R, Hudson. J, Pope. HG Jr.; Anabolic Androgenic Steroid Dependance: An Emerging Disorder, Addiction, 2009, Vol.104, Issue 12, pg. 1966-1978.

Keynes. R. D, Aidley. D.J, Huang. C.L-H; Nerve and Muscle, 2011, 4th ed. Cambridge; New York: Cambridge University Press, pg.99.

Khimji. D, Capadia. K; K11 Sports Nutrition Manual, 2011, accessed: https://books.google.co.uk/books?id=K6fHRvwq98QC&pg=PA8&dq=high+muscle+mass+metabolism&hl=en&sa=X&ei=3Zq3VOr1ApfbaofNgNAK&ved=0CCAQ6AEwAA#v=onepage&q=high%20muscle%20 mass%20metabolism&f=false.

Kicman. A.T; Pharmacology of Anabolic Steroids, British Journal of Pharmacology, 2008, Vol.154, Issue: 3, pg. 502-521.

Kinirons. M.T; Muscular Hypertrophy, French's index of differential diagnosis: an A-Z, 2011, CRC Press.

Kommi. P; The Encyclopaedia of Sports Medicine: An IOC Medical Commission Publication, Strength and Power in Sport, 2008, UK, Blakwell Science Ltd.

Korenman S, Wilson H, Lipsett M; Testosterone production rates in normal adults, Journal of clinical investigation, 1963, Vol.42, No.11.

Kuhn. C.M; Anabolic Steroids, Endocrine Society, 2002, Vol. 57, Ch. 19.

Kvorning. T, Kadi. F, Schierling. P, Andersen. M, Brixen. K, Suetta. C, Madsen. K; The activity of satellite cells and myonuclei following 8 weeks of strength training in young men with suppressed testosterone levels, Acta Physiologica, 2014.

Labire. F, Luu-The V, Lin. S.X, Labrie. C, Simard. J, Breton. R, Belanger. A; The key role of 17 beta- hydroxysteroid dehydrogenases in sex steroid biology, 2nd International Symposium on Molecular Steroidogenesis, 1997, Vol.62, Issue: 1, pg. 148-158.

Lamb DR; Anabolic steroids in athletics: how well do they work and how dangerous are they? The American Journal of Sports Medicine, 1984, 12(1):31-38.

Lass. A, Zimmermann. R, Oberer. M, Zechner. R; Lypolysis – a Highly Regulated Multi- Enzyme Complex Mediates the Catabolism of Cellular Fat Stores, Progress in Lipid Research, 2011, Vol.50, Issue: 1, pg.14-27.

Li. J, Al-Azzawi. F; Mechanism of androgen receptor action, Maturitas, 2009, Vol. 63, Issue:2, pg. 142- 8.

Loe. Y-c.C; Effects of Adipogenesis on Insulin Sensitivity: Two Time-course Studies in Diet-Induced Obese Mouse models, 2008, USA, ProQuest LLC.

Marquardt. G.H, Logan. C.E, Tomhave. W.G, Dowben. R.M; Failure of non-17-alkylated anabolic steroids to produce abnormal liver function tests, Journal of Clinical Endocrinology and Metabolism, 1964, *Vol.* 24, *pg.*1334–6.

Maughan. R.J, Watson. J.S, Weir. J; Strength and cross-sectional area of human skeletal muscle, Journal of Physiology ,1983, Vol.338, Issue:1, pg. 37-49.

Mebratu. Y, Tesfaigzi. Y; How ERK1/2 Activation Controls Cell Proliferation and Cell Death: is Subcellular Localization the Answer? Cell Cycle, 2009, Vol.8, Issue: 8.

Mohan. G; The Effects of Androgens on Cognitive, Morphological, and Neurochemical Functions in Female Rats, 2008, USA, ProQuest LLC

Mootz. RD, McCarthy.KA; Sports Chiropractic, 1999, USA, Aspen Publications, pg.168.

Morgan. J. E, Partridge. T. A; Muscle Satellite Cells, The International Journal of Biochemistry and Cell Biology, 2003, Vol.35, Issue: 8, pg.1151-1156.

Mottram D, George A; Anabolic steroids, Bailliere's Clinical Endocrinology and Metabolism, 2000, Vol.14, No.1, pg. 55-69.

Mougios. V; Exercise Biochemistry, USA, Human Kinetics, 2006, pg. 247.

Mozdziak. P.E, Schultz. E, Cassens. R.G; Myonuclear accretion is a major determinant of avian skeletal muscle growth, American Journal of Physiology – Cell Physiology, 1997, Vol.272, Issue: 2.

Nieschlag. E, Behre. H.M, Nieschlag. S; Testosterone: action, deficiency, substitution, 4th ed, 2012, UK, Cambridge University Press.

Ordway. G.A, Garry. D.J; Myoglobin: an essential hemoprotein in striated muscle, the Journal of Experimental Biology, 2004, Vol.207, pg. 3441-3446.

Osorio. M.S, Rojo. A.D, Benitez. B.M, Torre. A, Uribe. M; Anabolic Androgenic Steroids and Liver Injury, Liver International, 2008, Vol. 28, Issue: 2, pg.278.282.

Palvimo JJ. The androgen receptor. Mol Cell Endocrinol. 2012;352(1-2):1-3.

Pavlath. G; Myogenesis, 1st ed. 2011, USA, Elsvevier.

Payne. A.H, Youngblood. G. L; Regulation of Expression of Steroidogenic. Enzymes in Leydig Cells, Biology of Reproduction, 1995, Vol.52, Issue: 2, pg. 217-225.

Pi. M, Parrill. A.L, Quarles. L.D; GPRC6A Mediates the Non-genomic Effects of Steroids, The Journal of Biological Chemistry, 2010, Vol. 285, Issue:51.

Pope H, Kanayama G; Anabolic- Androgenic Steroids, Drug Abuse and Addiction in Medical Ilness, 2012, pg.251-264.

Pope. HG Jr., Kanayama. G; Anabolic Steroid Use: A Potential Public Health Problem?, Harvard Medical School, USA, 2012.

Porkka- Heiskanen. T, Kalinchuk. A, Alanko. L, Urrila. A, Stenberg. D, Adenosine, energy metabolism, and sleep, The Scientific World Journal, 2003, Vol.3, pg.790-8.

Powers. S.K, Talbert. E.E, Adhihetty. P.J; Reactive oxygen and nitrogen species as intracellular signals in skeletal muscle, The Journal of Physiology, 2011, Vol.589, Issue: 9, pg. 2129-2138.

Rainer. N. NG; The Occurrence of Contraction-Induced Lesions in the Sarcolemma of Skeletal Muscles: Insights from a Microsized Whole Muscle Model, 2008, USA, ProQuest LLC.

Ranjan. R, Parmar. A, Pattanayak. RD, Dhawan. A; Dependance on AAS: a case report and brief review, Delhi Psychiatry Journal, 2014, Vol.17, No.2.

Riebe. D, Fernhall. B, Thompson. P.D; The blood pressure response to exercise in anabolic steroid users, Med Sci Sports Exerc, 1992, Vol.24, Issue: 6, pg.633-7.

Riezzo I, De Carlo D, Neri M, Nieddu A, Turillazzi E, Fineschi V. Heart disease induced by AAS abuse, using experimental mice/rats models and the role of physical exercise. Mini Rev Med Chem. 2011, 11 (5): 409-424.

Rizzo S, Lodder EM, Verkerk AO, Wolswinkel R, Beekman L, Pilichou K, et al. Intercalated disc abnormalities, reduced Na(+) current density, and conduction slowing in desmoglein-2 mutant mice prior to cardiomyopathic changes. Cardiovasc Res. 2012;95(4):409.

Saartok T, Dahlberg E, Gustafsson JA. Relative binding affinity of anabolic-androgenic steroids: comparison of the binding to the androgen receptors in skeletal muscle and in prostate, as well as to sex hormone-binding globulin. Endocrinology. 1984;114(6):2100-6.

Schanzer W; Metabolism of anabolic androgenic steroids, Clinical Chemistry, 1996, Vol.42, No. 7, pg. 1001-1020.

Schiavi R, White D, Mandeli J; Pituitary-gonadal function during sleep in healthy aging men, Psychoneuroendocrinology, 1992, Vol. 17, Issue 6, pg. 599-609

Schmitt. J.M, Wayman. G.A, Nozaki. N, Soderling. T.R; Calcium activation of ERK mediated by calmodulin kinase I, the Journal of Biological Chemistry, 2004, Vol.279, Issue: 23, pg. 24064 – 72.

Shenkman. B.S, Turtikova. O.V, Nemirovskaya. T. L, Grigoriev. A. I; Skeletal Muscle Activity and the Fate of Myonuclei, ActaNaturae, 2010, Vol.2, No.2, pg. 59-66.

Sinha-Hikim. I, Roth. S.M, Lee M.I, Bhasin. S; Testosterone-induced muscle hypertrophy is associated with an increase in satellite cell number in healthy, young men, American Journal of Physiology - Endocrinology And Metabolism, 2003, Vol.285, Issue:1.

Sinha-H, Taylor. W.E, Gonzalez-Cadavid. N. F, Zheng. W, Bhasin. S; Androgen receptor in human skeletal muscle and cultured muscle satellite cells: up-regulation by androgen treatment, The Journal of Clinical Endocrinology and Metabolism, 2004, Vol. 89, Issue: 10, pg. 5245-5255.

Spiering. B.A, Kraemer. W.J, Vingren. J. L, Ratamess. N.A, Anderson. J.M, Armstrong. L.E, Nindl. B.C, Volek. J.S, Hakkinen. K, Maresh. C.M; Elevated endogenous testosterone concentrations potentiate muscle androgen receptor responses to resistance exercise, The Journal of Steroid Biochemistry and Molecular Biology, 2009, Vol.114, Issues:3-5, pg. 195-199.

Southren AL, Gordon GG, Tochimoto S, Pinzon G, Lane DR, Stypulkowski W; "Mean plasma concentration, metabolic clearance and basal plasma production rates of testosterone in normal young men and women using a constant infusion procedure: effect of time of day and plasma concentration on the metabolic clearance rate of testosterone". *J. Clin. Endocrinol. Metab.* 1967, 27 (5): pg. 686–94.

Southren AL, Tochimoto S, Carmody NC, Isurugi K ;"Plasma production rates of testosterone in normal adult men and women and in patients with the syndrome of feminizing testes". *J. Clin. Endocrinol. Metab.* 1965, 25 (11): 1441–50

Sriram D, Yogeeswari P; Medicinal Chemistry, 2nd edition, 2010, ch.34, pg.437

Structure of a muscle fibre; Image source:
http://faculty.weber.edu/nokazaki/Human_Biology/Chp%206-muscular-system.htm.

Sterngass. J; Steroids, 2010, USA, Marshall Cavendish Corporation.

Swash. M, Schwartz. M.S; Implications of Longitudinal Muscle Fibre Splitting in Neurogenic and Myopathic Disorders, Journal of Neurology, Neurosurgery and Psychiatry, 1977, Vol.40, Issue: 12, pg. 1152-1159.

Tan MHE, Li J, Xu HE, Melcher K, Yong E. Androgen receptor: structure, role in prostate cancer and drug discovery. Acta Pharmacol Sin. 2015; 2014;36(1):3.

Telley. I.A, Denoth. J; Sarcomere dynamics during muscular contraction and their implications to muscle function, Journal of Muscle Research and Cell Motility, 2007, Vol.28, Issue:1, pg.89-104.

Tesch. P.A, Larsson. L; Muscle hypertrophy in bodybuilders, European Journal of Applied Physiology and Occupational Physiology, 1982, Vol.49, Issue: 3, pg.301-306.

Transverse tubule, Encyclopedia of Molecular Pharmacology, 2008, pg.1242.

van Amsterdam. J, Opperhuizen. A, Hartgens. F; Adverse health effects of anabolic–androgenic steroids, Regulatory Toxicology and Pharmacology, 2010, Vol.57, Issue:1, pg.117-23.

van Marken Lichtenbelt, W.D, Hartgens. F, Vollaard. N.B.J, Ebbing. S and Kuipers. H; Bodybuilders' Body Composition: Effect of Nandrolone Decanoate. Med. Sci. Sports Exerc., 2004, Vol. 36, No. 3, pg. 484–489.

Verdijk. L.B; Satellite cell activation as a critical step in skeletal muscle plasticity, Experimental Physiology, 2014, Vol.99, Issue: 11, pg. 1449- 1450.

Vingren J, Kraemer W, Ratamess N, Anderson J, Volek J, Maresh C; Testosterone physiology in resistance exercise and training, Sports Medicine, 2010, Vol. 40, issue 12, pg. 1037-1053.

Wang. J, Conboy. I; Embryonic vs. Adult Myogenesis: Challenging the 'Regeneration Recapitulates Development' Paradigm, Journal of Molecular Cell Biology, 2010, Vol.2, Issue: 1, pg. 1-4.

West. D.W.D, Phillips. S.M; Associations of exercise- induced hormone profiles and gains in strength and hypertrophy in a large cohort after weight training, European Journal of Applied Physiology, 2011, Vol.112, Issue: 7, pg. 2693- 2702.

Williams. P; Plyometrics or not?, Strength and conditioning journal, 2001, Vol. 23, Issue: 2, pg. 70.

Yeap. B.B, Wilce. J.A, Leedman. P.J; The androgen receptor mRNA, BioEssays, 2004,Vol.26, Issue: 6, pg. 672-682.

I want morebooks!

Buy your books fast and straightforward online - at one of the world's fastest growing online book stores! Environmentally sound due to Print-on-Demand technologies.

Buy your books online at
www.get-morebooks.com

Kaufen Sie Ihre Bücher schnell und unkompliziert online – auf einer der am schnellsten wachsenden Buchhandelsplattformen weltweit!
Dank Print-On-Demand umwelt- und ressourcenschonend produziert.

Bücher schneller online kaufen
www.morebooks.de

OmniScriptum Marketing DEU GmbH
Heinrich-Böcking-Str. 6-8
D - 66121 Saarbrücken
Telefax: +49 681 93 81 567-9

info@omniscriptum.com
www.omniscriptum.com

www.ingramcontent.com/pod-product-compliance
Lightning Source LLC
Chambersburg PA
CBHW031541210526
45464CB00003B/1098